# A Guide to the Literature
## of Electrical and
## Electronics Engineering

# A Guide to the Literature of Electrical and Electronics Engineering

**Susan B. Ardis**
Edited by
Jean M. Poland

**1987**

**Libraries Unlimited, Inc.**          **Littleton, Colorado**

LIBRARIES UNLIMITED, INC.
P.O. Box 263
Littleton, Colorado 80160-0263

**Library of Congress Cataloging-in-Publication Data**

Ardis, Susan.
  A guide to the literature of electrical and
electronics engineering.

  (Reference sources in science and technology series)
  Includes index.
  1. Electric engineering--Bibliography.  2. Elec-
tronics--Bibliography.  I. Poland, Jean M.  II. Title.
III. Series.
Z5832.A83  1987    [TK145]    016.6213    87-25981
ISBN 0-87287-474-5

Libraries Unlimited books are bound with Type II nonwoven material that meets and exceeds National Association of State Textbook Administrators' Type II nonwoven material specifications Class A through E.

# Contents

### Part 1
### Bibliographic Sources

## Part 2
### Ready Reference Sources

## Part 3
### Handbooks and Reference Texts

## Part 4
### Journals and Newsletters

## Part 5
### Product Literature

**Part 6**
**Patents**

**Part 7**
**Standards**

**Part 8**
**Other Information Sources**

# Preface

This book is intended for technical library users, who may be students, reference librarians, or others seeking specific information. It is meant to serve as a guide to the modern literature of electrotechnology. *Electrotechnology* is used as a general rubric for a number of interrelated fields concerned with the use and study of electrical energy, whether in large amounts at power stations or in small amounts in circuits. Areas covered in electrotechnology include electrical and electronics engineering, with the traditional subdivisions of power engineering, communications, circuits and systems, control systems, electron and solid-state devices, signal processing, and computer engineering. Selected materials from related areas such as microwaves, radar, materials science, and satellite technology have been included because electrical and electronics engineering is not an independent area. Areas generally excluded include robotics, CAD/CAM, appliance repair, electrical wiring, and electrical construction. Some areas specifically excluded or given limited coverage include nuclear power, fossil fuels, alternative energy, and public policy.

Coverage of physics and mathematics is limited to those books which relate directly to the design, development, utilization, application, or manufacture of electrical or electronic devices. The same selection criteria were used for software or programming literature. In the area of electrotechnology, only those books and guides which provide direct access to software are included. Communications is another traditional area of electrical engineering that has been affected by electronic devices and computer technology. Those areas of communications that do not relate directly to either signal processing or the circuitry used in communicating over air waves or cables are specifically excluded from this guide. Applications of specific electronic devices in musical instruments, biomedical applications, and avionics are excluded except where they are part of other materials.

No guide to the literature ever claims to be comprehensive, and this book is no exception. Comprehensiveness is especially difficult to achieve in a dynamic and competitive area such as electrotechnology, which regularly incorporates new inventions and developments into its body of knowledge. This new knowledge must then be assimilated and transmitted to thousands of students, researchers, and engineers in the field. The need to transfer information from researchers to students, designers, and manufacturers has not gone unnoticed by publishers, who have responded by publishing thousands of books each year. Of the materials available, the majority of those chosen for inclusion here were published since 1978. Some that were not in-print were selected because they contain valuable information, are still popular with users, or are classics. To be selected for inclusion, a title had to be in English, or abstracted or translated into English.

Basic reference materials and general library tools were excluded, as these are covered in other bibliographies. Of those materials selected, over 95 percent were examined by the author. The materials were selected from the author's library, Texas A&M's library, publishers' catalogs, and OCLC. Every attempt has been made to be as current and accurate as possible.

Not only does this guide seek to be as up-to-date as possible, but it also seeks to reflect the variety and abundance of relevant reference materials. The term *reference* is used quite broadly to mean any book which facilitates "looking up" information or data. Coverage includes traditional reference materials such as handbooks, indexes, guides, and dictionaries, along with the more specialized or unusual reference sources such as data compilations, standards, patents, product literature, and newsletters. Emphasis has been placed on these special reference materials because of the nature of electrotechnology, a field whose goal is to design, fabricate, and manufacture products.

A number of intellectual criteria besides language, currency, and status were used to select materials for inclusion. The book must provide easy access to the material covered through indexes and tables of contents, allow quick consultation, and provide sources for the data or information presented. The text should be divided into relevant sections, with tables, formulas, and graphs logically arranged, clearly labeled, and indexed.

Some areas are represented here in more depth than others, and some areas are specifically excluded. This reflects the extent of publishing in each category. An unplanned emphasis of the book is on electronics, including digital and solid-state devices. This is in keeping with current developments. The decline of power engineering is clearly demonstrated by the small number of materials published since 1978, and those that have appeared are heavily influenced by digital solid-state electronics.

The primary arrangement for the book is by type of material (product catalogs, dictionaries, etc.). Some of the larger categories, such as handbooks, are further subdivided by subject to facilitate use. If a book could be listed in more than one section, its location was determined by the primary focus of the book. Entries are arranged in title order, because many of the books are compilations of a number of authors and the titles of the books are descriptive. Each entry gives full bibliographic information and a short descriptive annotation. Imprints include place, publisher, and date, and are as complete as possible. Pagination includes only Arabic numbered pages in the main body of the book. The illustration statement encompasses graphs, photographs, drawings, schematics, charts, etc. Individual types of illustrations are listed in the annotation if they make an important contribution to the book. ISBN, ISSN, LC, SuDoc, and NTIS numbers are included when available from the piece or from a search of *Books in Print,* the *Monthly Catalog,* or *Government Reports Announcements.* Most of the ISBNs were taken from the piece itself. Price is included when available. (Listed prices should be used only as an indication of the relative cost.)

One of the most commonly asked questions by new librarians or collection development personnel is "who are the most important publishers in your field?" This is often not easy to answer and the answer is generally based on anecdotal evidence. For this reason the following chart has been included. This chart lists all book publishers with more than three listings in this guide.

| Publisher | Number of Entries |
|---|---|
| McGraw-Hill | 56 |
| Prentice-Hall | 41 |
| Wiley, Wiley-Interscience Publication | 32 |
| D.A.T.A. Inc. | 27 |
| Tab Books | 25 |
| Van Nostrand Reinhold | 19 |
| Howard W. Sams | 16 |
| Reston Publishing | 16 |
| IEEE Press | 9 |
| Elsevier, Elsevier Science Publishing | 9 |
| IEE | 7 |
| Butterworth Publishers | 6 |
| Artech House | 5 |
| Hayden Publishing | 5 |
| Pergamon Press | 5 |
| Heyden | 4 |
| Plenum | 4 |
| Academic Press | 4 |
| Addison-Wesley | 3 |
| Bowker | 3 |
| Information Handling Service | 3 |
| Marcel Dekker | 3 |
| NTIS | 3 |
| Research Publications | 3 |
| Publishers with less than 3 | 88 |
| | 396 |

# Introduction

The literature of electrotechnology is large and varied, incorporating theoretical and practical information from a number of related areas. It encompasses, for example, theoretical information on crystal lattice structure as well as practical information on the manufacture and application of electronic devices. The tools used to access the literature reflect this dichotomy.

Finding information in electrotechnology requires knowledge of the field, understanding of the intellectual organization of the subject, and familiarity with the tools. The first step in looking for information is to become acquainted with the field and its tools.

A simple question to ask in determining where to look for specific information is "what type of engineer would be interested in the type of information I am seeking? Would it be a researcher, a designer, a manufacturer, a fabricator, or a technician in electrical, electronics, or computer engineering?" The answer helps determine where the information being sought would be found. Practical information, such as the description of a particular chip, is found in tools such as manufacturers' catalogs or datasheets. Theoretical information, such as lattice structure, is found in handbooks and data compilation which access theoretical materials.

The following short definitions of major components of electrotechnology are included to provide a quick introduction to the field. Electrical, electronic, and computer engineering differ from the basic sciences in a substantive way. Basically the difference centers around the nature of the work. Generally, engineering is a task-oriented field centering around the application of existing principles to the design and development of products, whereas the task of the basic scientist is to discover these principles.

There is still substantial overlap between the basic sciences and engineering. This overlap is the strongest between physics and electrical engineering, which share a common history and interest in the physical phenomena of the electric charge. This is the basis for all electrical, electronic, and computer engineering. The earliest recorded observation of the electrical effect was made by the Greeks, who observed that when amber was rubbed it attracted straw.

Over the centuries, advances in electrical engineering have been closely allied with a number of well-known inventions and discoveries. This inheritance from the basic sciences and from engineering has been acknowledged through naming units or ideas after the early investigators and scientists. The *volt,* the *faraday*, the *watt,* the *ohm,* and the Edison Effect are all named after men who advanced the study and use of electricity.

Traditionally, electrical engineering has been concerned with the generation, transmission, distribution, utilization, application, and control of electrical power in large systems. But this interest went deeper than the system itself. It also included the components of those systems which, when wired together, performed a specific task such as generating power. System components are integral to electrical engineering; they serve two basic functions. The components serve either as switching or as regulating mechanisms. The switching (digital) component has two positions, ON or OFF. The regulating (analog) component controls the flow of electrons through a continuous state. An example of a simple switching/digital device is the doorbell. The radio volume control serves as an example of a regulating/analog device. The development and improvement of digital devices is having a profound impact on electrical engineering and leads directly to electronics. The study of electronics, like electrical engineering, is the result of basic discoveries in physics. The goal of the early physicists was to understand electricity, the goal of electrical engineers is to control electricity, and the goal of electronics is to control the electron. Control at first was accomplished through the use of mechanical switches. The invention of the electron tube dramatically changed electrical engineering by allowing the switching and regulating functions to take place at the speed of electrons or electronically.

One of the simplest explanations of the difference between electrical and electronic engineering involves the level of energy used. Electronics deals with lower energy levels; it is concerned with the release, transport, control, collection, and conversion of elementary charges. Over the last 60 years the development of electronics has passed through three distinct phases. Each phase has corresponded roughly to the invention of a major active device: the thermonic tube or valve, the transistor, and the integrated circuit. The invention in 1948 of the solid state transistor had a dramatic effect on the growth of electronics. The transistor allowed the combination of formerly separate active and passive components into the integrated circuit, or IC. The invention of the IC revolutionized electronics and ushered in the development of the digital computer.

Of all the areas associated with electrical engineering, none is advancing as rapidly as computer engineering. Computer engineering, while related to computer science, involves the circuits and design of the computer. Computer engineering can be broken down into three areas: analog or hybrid computation, digital computing, and computer applications. Digital computing has been influenced most directly by electrical and electronics engineers. Digital computing may be subdivided into system description and design, logic systems, circuit design, and storage systems. Within these divisions it is system design and circuit design in which engineers have had the greatest impact.

Over the centuries, electrical engineering has advanced from attempts to understand the basic phenomenon of electricity to the use of electronic devices in thousands of applications, including communications and computers. Each new invention or advance in knowledge has led to the creation of new products, and these new products have led to the creation of new fabrication and manufacturing techniques. These inventions created the modern electronic industries which have changed the lives and careers of everyone from office workers to engineers. Thus, the impact and importance of the distribution, control, and utilization of the electric phenomenon cannot be overstated.

# History

The historical development of electrical and electronics engineering is the study of early physicists and their drive to understand the phenomena of electricity and magnetism. This history also reflects the endeavor of inventors and early engineers to harness the power of electricity in order to drive machines and to communicate over long distances faster, more efficiently, and at lower cost.

**An Age of Innovation: The World of Electronics, 1930-2000.** Edited by *Electronics Magazine*. New York: McGraw-Hill, 1981. 274p. illus. index. LC 80-14816. ISBN 0-07-606688-6. $18.50.

This special 50th anniversary issue of *Electronics Magazine* describes developments during the period 1930-1980. The first section covers the discovery of the Edison Effect through the first superheterodyne receiver, while the second section deals with the radio years and "electronics at war." The development of transistors, computers, and the digital age are discussed in the last section. The many illustrations include classic circuits such as the flip-flop. The volume also has short biographies and pictures of many of the inventors and scientists who contributed to the development of electronics.

**Bibliography of the History of Electronics.** By George Shiers. Metuchen, N.J.: Scarecrow Press, 1972. 336p. index. LC 72-3740. ISBN 0-8108-0499-9.

A collection of 1,820 descriptive annotations of books, articles, and reports on the history of electronics and telecommunications from the 1860s to the 1950s.

**Electrical and Electronic Technologies: A Chronology of Events and Inventors to 1900.**
By Henry B. O. Davis. Metuchen, N.J.: Scarecrow Press, 1981. 221p. index. LC
81-9179. ISBN 0-8108-1464-1. $16.00.

**Electrical and Electronic Technologies: A Chronology of Events and Inventors from
1900-1940.** By Henry B. O. Davis. Metuchen, N.J.: Scarecrow Press, 1983. 220p. index.
LC 82-16739. ISBN 0-8108-1590-7. $16.00.

**Electrical and Electronic Technologies: A Chronology of Events and Inventors from
1940-1980.** By Henry B. O. Davis. Metuchen, N.J.: Scarecrow Press, 1985. 321p. index.
LC 84-13957. ISBN 0-8108-1726-8. $25.00.

A three-volume chronology of important inventors and inventions in electrical and
electronics engineering. Divided by decades, each division is prefaced by a five- or six-
page introduction and summary of the events that occurred during that decade.
Individual events are then listed in chronological order by year within each decade.
Thus one has a year-by-year account of both major and minor events. An event may
also be listed by concept, such as "osmium lamp" or by the individual responsible for
the event. Each volume has a separate index.

**Electricity in the 17th & 18th Centuries: A Study of Early Modern Physics.** By J. L.
Heilbron. Berkeley, Calif.: University of California Press, 1979. 606p. index. LC
77-76185. ISBN 0-520-03478-3. $60.00.

A scholarly treatment of the early history of physics as it relates to the understanding and development of commercial electricity. It includes an extensive bibliography of 68 pages which cites both published works and manuscripts. The first section
recounts the development of electricity in the 17th century. In the second section, the
great discoveries like the ubiquity of electricity, the age of Franklin, and the quantification of electricity are described.

**The Electronic Epoch.** By Elizabeth Antebi. New York: Van Nostrand Reinhold, 1983.
280p. illus. index. LC 82-51015. ISBN 0-442-28254-0. $49.50.

Large book with extensive color photographs and drawings which gives an explanation of the history, the importance, and the use of electronics. Examples of important
breakthroughs are used to illustrate some of the concepts. The first part discusses the
nature of electronics while the rest of the book covers major inventions like the
transistor and how they affected later electronic developments. An extensive bibliography is included as well as subject and name indexes.

**Electronic Inventions and Discoveries.** 3rd ed. By G. W. A. Dummer. Oxford:
Pergamon Press, 1983. 220p. index. LC 83-2393. ISBN 0-80-029354-9. $44.00.

Discusses the developmental history of important electronic inventions and discoveries including resistors, capacitors, valves, transistors, and integrated circuits. The
author believes the development of these components is the story of electronics. Details
of 500 such inventions are covered. It is interesting to note how many of these electronic
inventions resulted in the development of sizable industries.

A portion of the book consists of lists; lists of patents related to each other, lists of
inventions, and lists of inventions by subject. However, most of the book is devoted to
brief descriptions, in chronological order, of 500 major inventions from 1642-1982. A
short list of references is given after each invention.

**The Engineering Profession: Its Heritage and Its Emerging Public Purpose.** By Dan H. Pletta. Washington, D.C.: University Press of America, 1984. 262p. illus. index. ISBN 0-8191-3835-5. $27.50.

Primarily a guide to the career of engineering, this book includes the history of engineering. Of special interest is the discussion of the philosophical and ethical goals of the profession.

**Engineers & Electronics: A Century of Electrical Progress.** By John D. Ryder and Donald G. Fink. New York: IEEE Press, 1984. 251p. illus. index. LC 83-22681. ISBN 0-87942-172-x. $29.95.

A beautifully illustrated history of electrical engineering in the United States presented from the perspective of the IEEE and the earlier related professional societies. The book describes the discovery of electricity, the development of electrical power distribution, and the rise of professionalism in electrical engineering.

**From Spark to Satellite: A History of Radio Communication.** By Stanley Leinwoll. New York: Scribners, 1979. 241p. illus. index. LC 78-24172. ISBN 0-684-16048-x.

A history of wireless communication from Michael Faraday's early electrical experiments in the 1820s, to the theoretical work of James Maxwell, and the patents fights for commercial rights to the wireless. The text details the commercialization of radio and ends with a discussion of the future of lasers and satellites in communications. Includes a bibliography.

**History of Electric Light and Power.** By B. Bowers. London: Peter Peregrinus, 1982. 304p. illus. index. ISBN 0-906048-68-0. $80.00. (IEE History of Technology Series, number 3).

Describes the origins and growth of the supply industry and the development of the primary uses of electricity. European developments are stressed.

**History of Electric Wires and Cables.** By R. M. Black. London: Peter Peregrinus, 1983. 340p. illus. index. ISBN 0-86341-001-4. $60.00. (IEE History of Technology Series, number 4).

The evolution of electric wires and cables from those used in early telegraphy to modern examples like optical fiber and telecommunication cables is covered. Also includes a discussion of the principal types of electric wires and cables used during the late 1950s.

**History of Electrical Power Engineering.** By Percy Dunsheath. Cambridge, Mass.: MIT Press, 1969. 368p. illus. index. $9.95.

Topics include the birth of electric telegraphy, Michael Faraday's contribution, electricity supply, AC current, the early telephone, and the development of professional organizations. A chronological table of social and historical background information is included. Photographs illustrate early equipment like a 1920s electric power station, an early switchboard, and a 1908 electrical locomotive. A bibliography is included.

**Pioneers of Electrical Communication.** By Rollo Appleyard. Freeport, N.Y.: Books for Libraries Press, c1930, 1968. 347p. illus. index. LC 68-54322. ISBN 0-8369-0156-8. $24.50.

Profiles of 10 men who contributed to the advancement of electrical communication by helping to bridge the gap between electricity as a branch of physics and electricity as an engineering science. Among those profiled are James Maxwell, Charles Wheatstone, Hans Oersted, and Benjamin Franklin.

**Revolution in Miniature: History and Impact of Semiconductor Electronics Re-explored.** 2nd ed. By Ernest Braun and Stuart Macdonald. London: Cambridge University Press, 1982. 247p. index. LC 82-1117. ISBN 0-521-24701-2. $32.50.

The history and impact of the semiconductor on individuals and on industry are covered. Aimed at the intelligent layman, the book gives a very complete account of the development of miniature electronics. Only minor changes were made to the first eight chapters in the Second Edition. However, chapters 9, 10, and 12 which cover very large scale integration, the American semiconductor industry, and a reflection on the Electronics Age were heavily revised and updated. An extensive list of primary and secondary references is included. Many references are to personal communications with individuals who made important contributions to the development of semiconductor electronics.

**Syntony and Spark: The Origins of Radio.** By Hugh G. J. Aikten. Princeton, N.J.: Princeton University Press, 1985. 368p. illus. index. LC 84-26408. ISBN 0-691-08377-0. $38.50.

Details the theory and technology of electricity and radio from the work of J. Clerk Maxwell in the 1860s to that of Marconi in the early 20th century. Offers an especially strong analysis of the relationships among science, technology, and society. Extensive references are provided at the end of most sections.

**Volts to Hertz: The Rise of Electricity.** By Sanford P. Bordeau. Minneapolis, Minn.: Burgess Publishing, 1982. 308p. illus. LC 82-17702. ISBN 0-8087-4908-0. $18.95.

Advances in electricity are examined through the stories of 16 men who contributed to the field. The text is illustrated with portraits of each man. Each biography includes family background and education as well as discussion of their achievements. Some of the men covered are William Gilbert, James C. Maxwell, and Nikola Tesla. Contains references and an index of names.

**PART 1**
Bibliographic
Sources

# 1   Guides to the Literature

The guides listed in this section are neither abstracting nor indexing services. Rather they are bibliographic lists describing the secondary sources of information in science and technology designed to help the user find his or her way through the different types of literature much as a map or guidebook helps the traveler. These sources may be aimed at students, working engineers, or librarians. Many will also include helpful hints on research techniques, using the library, or organizing research notebooks. Because there are no current guides covering electrical and electronics engineering specifically, the following deal with science and engineering in general. Most have specific sections on various disciplines.

1.   **Abstracts and Indexes in Science and Technology: A Descriptive Guide.** 2nd ed. By Dolores B. Owen. Metuchen, N.J.: Scarecrow Press, 1985. 235p. LC 84-10902. ISBN 0-8108-1712-8. $17.50.

The aim of this guide is to provide a description of abstracts and indexes arranged by subject. Each annotation consists of the following information: arrangement, coverage, scope, locating materials, abstracts, indexes, other materials, and database.

2. **Finding Answers in Science and Technology.** By Alice L. Lefler Primack. New York: Van Nostrand Reinhold, 1983. 364p. index. ISBN 0-442-28227-3. $22.50.

Aimed at all levels of user, from student to librarian, this literature guide covers the areas of computer science, engineering, mathematics, astronomy, physics, and chemistry. Of special interest are the chapters devoted to online searching, which is aimed at the nonlibrarian, and the listing of libraries which have more than 500,000 volumes in science and technology. The author is the physics and engineering librarian at the University of Florida. All scientific collections would benefit in owning this particular title.

3. **How to Find Out about Engineering.** By S. A. J. Parsons. Oxford: Pergamon Press, 1972. 285p. index. LC 72-83292. ISBN 0-08-016919-8. $25.00.

A guide to sources of information on engineering and its subdivisions, with a heavy bias toward European sources. Has a good overview of the large engineering societies. While most of the cited handbooks and dictionaries are now dated, the search techniques covered are still useful.

4. **How to Find Out in Electrical Engineering: A Guide to Sources of Information Arranged According to the Universal Decimal Classification.** By Jack Burkett and Philip Plumb. Oxford: Pergamon Press, 1967. 234p. illus. index.

Although very dated in its coverage of electronics handbooks and databooks, this book does give the reader a good introduction to the general organization of the literature. The introductory chapter on careers, and the sections on the organization, acquisition, and selection of materials should interest librarians new to the field as information remains valid today.

5. **Information Resources for Engineers and Scientists; Workshop Notes.** 4th ed. Edited by Charlie Maiorana. White Plains, N.Y.: Knowledge Industries, 1985. 725p. looseleaf. index. ISBN 0-86729-189-3. $175.00.

Originally notes given to attendees at a conference designed to help scientists and engineers make more effective and efficient use of technical libraries, this guide has been expanded to be an overall treatment of technical information. Covers about 50 categories of reference sources, including technical reports, conference proceedings, guidebooks, U.S. documents, patent information, translations, and NTIS (National Technical Information Service). A typical section ha a few pages of general introduction followed by descriptions of specific reference works in the category.

6. **Information Sources: A Guide for the Engineer.** Edited by Terry Shoup. Englewood, Colo.: Information Handling Service, 1984. 48p. illus. index. ISBN 0-898-470099. $12.00.

A basic overview of problems and complexities of the engineering literature, this book is designed to help the engineer find information in books, journals, catalogs, government regulations, standards, and abstracts. Primarily a public relations piece on why engineers need to use more information tools, this book is useful as a training tool in the use of products from Information Handling Service.

7. **Information Sources in Science and Technology: A Practical Guide to Traditional and Online Sources.** 2nd ed. By C. C. Parker and R. V. Turley. London: Butterworth Publishers, 1986. 288p. ISBN 0-408-01467-9. $64.95.

A guide for practicing scientists and engineers in the United Kingdom. Because it is divided into sections by type of material, all listings for reference materials, government publications, and patents are together. Citations include only the city or country, no publishers are listed. Each section includes separate discussions of uses, access points, and cautionary notes.

8.   **Science and Engineering Literature: A Guide to Reference Sources.** 3rd ed. Edited by H. Robert Malinowsky and Jeanne M. Richardson. Littleton, Colo.: Libraries Unlimited, 1980. 342p. index. LC 80-21290. ISBN 0-87287-230-0. $33.00.

This book has consistently been used as a text for courses covering information sources in science and technology since its first edition in 1967. The annotations describe the scope of the work, intended audience, and special features of the work. The book lacks a separate section for either electrical or electronic engineering material. Materials on this subject are listed in the section with books on mechanical engineering, resulting in very light coverage of electrical and electronics engineering topics.

9.   **Scientific and Technical Information Sources.** 2nd ed. By Ching-Chih Chen. Cambridge, Mass.: MIT Press, 1986. 580p. index. LC 77-9557. ISBN 0-262-03120-5. $55.00.

The intended audience for this book is science and engineering librarians. Sources are categorized first by types of materials and then by subject within each type. Over 3,650 sources are grouped under 23 categories. Entries are arranged by title. Most of the annotations are descriptive rather than critical. A detailed table of contents and a complete author index are provided.

10.   **Scientific and Technical Information Sources.** By Krishna Subramanyam. New York: Marcel Dekker, 1981. 416p. index. LC 80-29531. ISBN 0-8247-1356-7. $28.95.

For students and teachers of librarianship and practicing librarians, this book is organized by types of materials. Over 1,500 sources are listed under broad categories, including tertiary sources like guides to the literature, secondary sources such as bibliographies, abstracts, and catalogs, and primary sources like journals and patents. Each of the broad sections is prefaced by a short expository essay describing the materials covered, the usefulness of the materials, etc. The annotations follow, arranged in subject order.

11.   **U.S. Access to Japanese Technical Literature: Electronics and Electrical Engineering.** Washington D.C.: GPO, 1986. 155p. SN 003-003-02709-0. $6.00. (National Bureau of Standards Special Publication 710).

The results of a seminar held in 1985 at NBS which was attended by representatives from congress, industry, universities, and federal agencies. Consists of ideas and techniques for obtaining access to the 80 percent of Japanese technical literature which is never translated.

12.   **Use of Engineering Literature.** Edited by K. W. Mildren. Woburn, Mass.: Butterworth Publishers, 1976. 621p. illus. index. ISBN 0-408-70714-3. $99.95.

A survey which chronicles the literature of electronics, communications, control engineering, aeronautics and astronautics, chemical engineering, production engineering, and soil engineering. Each chapter includes information on classification, indexing, journals, conferences, translations, reports, standards, and product information.

13.   **What Every Engineer Should Know about Engineering Information Resources.** By Margaret T. Schenk and James K. Webster. New York: Marcel Dekker, 1984. 213p. index. LC 84-11350. ISBN 0-8247-7244x. $28.00. (What Every Engineer Should Know Series, vol. 13).

One of a series of practical books directed at practicing engineers. Includes periodicals, nonbibliographic databases, bibliographic databases, handbooks, and patents among other topics. Of special interest is the small section on computer software for working engineers.

# 2 Bibliographies

This section consists of a very select list of bibliographies on topics of interest to researchers in electrotechnology. In the past, prepared bibliographies were an important source of information on highly specialized topics or for areas not well served by indexes or abstracts. However, the usefulness of bibliographies has faded somewhat with the introduction of bibliographic databases. While not as important or as popular as in the past, bibliographies still play an important role in accessing information prior to the middle 1960s and for information that is not indexed or included in abstracting services or databases. The following books are good examples of bibliographies which still meet a need as they consist of collections of materials not found in the major databases or material published prior to the databases.

14. **Bibliographic Guide to Technology**. Boston: G. K. Hall, 1976- . semiannual. ISSN 0360-2761. $250.00/yr.
   This guide consists of catalog records for materials in all languages and formats cataloged by the Library of Congress and Research Libraries of The New York Public Library along with conferences from the Engineering Societies Library. Includes complete LC, ISBN, NYPL, and Engineering Societies Library cataloging information for each title. Access is by main entry, added entries, titles, series titles, and subject headings.

15.   **Classed Subject Catalog of the Engineering Societies Library and Index to the Classed Subject.** 2nd ed. revised and updated. New York: Engineering Societies, 1984. 335,000 cards. 22 reels. index. ISBN 0-8161-1333-5. $2,600.00.

This significantly revised and updated second edition includes all entries from the first edition. Of interest are the library's extensive holdings of journals and conference proceedings. It is organized by the Universal Decimal Classification scheme and has a separate index volume.

The Engineering Societies Library owns copies of most of the materials abstracted by *Engineering Index.* The library is not associated with OCLC, RLIN, or any of the computer literature searching vendors. Therefore, access to materials held in the library is possible only through this catalog or by inspecting their onsite catalog. It is important to note that members of societies like the ASME, IEEE, or ASCE may borrow directly from the society; other institutions may order photocopies for a fee.

16.   **Handbooks and Tables in Science and Technology.** 2nd ed. Edited by Russell Powell. Phoenix, Ariz.: Oryx Press, 1983. 304p. index. LC 82-19842. ISBN 0-89774-039-4. $60.50.

An annotated compilation of handbooks and tables arranged in alphabetical order by main entry, this volume covers books from all areas of science and technology including electrical and electronics handbooks. Indexed by author/editor, broad subjects, and publishers, it would have been easier to use if the annotations had been grouped by subject.

17.   **Ion Implantation in Microelectronics: A Comprehensive Bibliography.** Compiled by A. H. Agajanian. New York: IFI/Plenum, 1981. 266p. index. LC 81-10753. ISBN 0-306-65198-2. $85.00. (Computer Science Information Guides, vol. 1).

Suitable for strong fabrication and electronic materials collections, this book provides in-depth coverage of the literature prior to 1980. This is a good starting point for students and researchers interested in the early literature of the growing field of ion implantation. The bibliography is not annotated but most references include the volume and abstract number for materials taken from *Electrical and Electronics Abstracts, Engineering Abstracts, Chemical Abstracts,* and *Books in Print.*

18.   **Japanese Journals in English.** By Betty Smith and Basil Alexander. London: British Library Lending Division and the British Library Science Reference Library, Longwood Publishing Group, 1985. 138p. ISBN 0-7123-0721-4. $18.00.

A collection of translated Japanese scientific, technical, and commercial journals published from 1960-1984 which are held by the British Library Science Reference Library and/or the British Library Lending Division. This publication is divided into three sections: a subject key word list of English language journals published in Japan, a key word list of translated journals and journals with translations, and a listing of the original title.

19.   **Microelectronic Packaging: A Bibliography.** Compiled by A. H. Agajanian. New York: IFI/Plenum, 1979. 254p. LC 79-18930. ISBN 0-306-65183-1. $85.00.

A bibliography covering information on the housings of electronic devices, which is a very important area of design work for those engaged in preparing designs for manufacturing. Most references are from *Chemical Abstracts, Books in Print,* and *Science Abstracts.*

20. **MOSFET Technologies: A Comprehensive Bibliography.** Compiled by A. H. Agajanian. New York: IFI/Plenum, 1980. 377p. index. LC 80-21773. ISBN 0-306-65193-9. $95.00. (IFI Data Library Series).

The metal oxide semiconductor field effect transistor, more commonly known as MOSFET, was invented in 1960. The processing technology required for its successful high-volume fabrication became available in 1968. This book contains a comprehensive bibliography of over 4,440 references in world literature to MOSFET, GaAs FET, silicide gate FETs, and VLSI. Along with the author's *Semiconducting Devices: A Bibliography of Fabrication Technology* (New York: IFI/Plenum, 1976), this book covers the literature from 1976 to 1980. The main sources were *Electrical and Electronics Abstracts, Chemical Abstracts, Engineering Index, NTIS, Books in Print, Cumulative Book Index*, current journals, and conference digests. The material is arranged by technologies, properties and characterizations, and dielectric thin films. The subject index is very comprehensive.

21. **Pure & Applied Science Books 1876-1982.** New York: R. R. Bowker, 1982. 7,139p. 6v. index. LC 82-12862. ISBN 0-8352-1437-0. $300.00.

Some 220,000 titles organized by Library of Congress subject headings are included. Each subject entry is a complete LC card including notes and tracings. The usefulness is increased by the addition of a number of form entries, such as dictionaries and encyclopedias. The author and title indexes contain an abbreviated entry with references back to the main subject entry. Volume 2 covers electrical engineering and electronics. The dates given in the title are slightly misleading as there are more than 3,500 entries prior to 1869. This is an excellent source of information about books on early engineering.

# 3   Abstracts and Indexes

Abstracts and indexes provide access to the periodical literature of many fields including electrical and electronic engineering, and the related areas of computer science and telecommunications. Because the literature in this area is so large and complex, there are several major indexes and abstracts which provide access to selected segments of the published literature. Many of the following indexes provide access not only to the traditional periodical literature but also selected technical reports, conference proceedings, patents, and product literature.

The basic difference between an index and an abstracts is in the amount of information provided. Indexes provide the basic bibliographic information: author's name, title of the article, title of the periodical, volume and page number, date of publication, and often the work address of the first author. Abstracts, on the other hand, provide the standard bibliographic information but add a short summary of the article.

In chapter 5 indexes are listed which are not limited to electrotechnology or even engineering, but which cover specialized types of materials like conferences and proceedings or technical reports. These have been included both as a reminder of their existence and because of their importance as finding aids in engineering.

22.   **Applied Science and Technology Index.** New York: H. W. Wilson, 1959- . monthly. ISSN 0003-6989. price varies. (Formerly *Industrial Arts Index*).

A subject index to about 225 English language periodicals covering both the theoretical sciences and their engineering applications. Provides substantial coverage of scientific applications, including electrical engineering, electronics, and computer science, as they are represented in the scientific periodicals covered. This is not a serious limitation for most users since coverage includes all the journals as well as transactions from the IEE and IEEE. Also included are the major trade journals associated with these fields. Cumulated quarterly and annually, this is a good basic index for libraries and individuals interested in the most generally available materials in science and technology. Book reviews are listed at the end of each cumulation. Available as file AST on WILSONLINE.

23.  **Computer and Control Abstracts.** Vol. 1- . London: INSPEC, 1966- . monthly. ISSN 0036-8113. $610.00/yr. (Science Abstracts Series C).

An important international abstracting journal covering systems and control theory; computer and control technology and applications; and systems and equipment. Over 24,000 items are abstracted each year. Heavy emphasis is given to literature from conferences, especially those held in Europe. All abstracts are in English with the language of the original article indicated. Issued monthly with annual indexes, coverage includes books, conferences, patents, and journals. The monthly issues are arranged by a hierarchical subject classification and include an author index. The subject classification is published separately on the back cover of each monthly issue. The lack of a subject index in each monthly issue is only a minor drawback, since all related articles indexed that month are found together. Cumulated author and subject indexes are available including the most recently made available, 1980-1984. They may be purchased separately or as a set. The price for the set is $925.00.

This index is part of the INSPEC indexing services which is the major abstracting and indexing service in physics and engineering. Other parts of the INSPEC service include *Physics Abstracts* and *Electrical and Electronics Abstracts* (see entry 27). While each section functions as a separate unit, there is substantial duplication between the parts, especially between this index and *Electrical and Electronics Abstracts*. A supplementary current awareness publication, *Current Papers on Computers and Control* (see entry 38) is also for sale. The computerized equivalent is INSPEC.

24.  **Computer and Information Systems Abstract Journal.** Vol. 1- . Riverdale, Md.: Cambridge Scientific Abstracts, 1962- . monthly. ISSN 0191-9776. $590.00/yr.

An international abstracting journal for material on computer software, computer applications, computer mathematics, and computer electronics. Lists periodicals, government reports, conferences, books, dissertations, and patents. Access is provided for acronyms, subjects, authors, and sources.

25.  **Computing Reviews.** Vol. 1- . New York: Association for Computing Machinery, 1960- . monthly. ISSN 0010-4884. $50.00/yr.

Provides short, signed reviews of computer literature in a classified arrangement. Emphasis is on machine aspects, such as hardware and software. However, there are also sections for textbooks, proceedings, and the mathematics of computation. Each volume has an author and cumulative permuted KWIC index (key word in context).

26.  **EI Engineering Conference Index.** Vol. 1 . New York: Engineering Information, 1984- . monthly. $695.00/yr. $155.00/ea.

This print version of *EI Engineering Meetings* provides access to published papers which were presented at engineering meetings. Entries consist of the bibliographic citations to over 100,000 papers published in more than 2,000 proceedings, symposia, conferences, and workshops held throughout world. Organized by discipline into six volumes. Of particular interest are volume 4, which covers electrical, power, optical, acoustical engineering, and applied optics, and volume 5, which deals with electronics, information science, communications, computer and control engineering, applied mathematics, and instrumentation. In 1984 20,700 papers from 310 conferences were indexed in volume 4 and 11,000 papers from 200 conferences in volume 5. Each volume has separate author, subject, author's affiliation, and conference title indexes. Also included is a cross-reference code to the *Engineering Index* abstract entry for the complete meeting. Beginning in 1985, each entry includes an abstract. Because *EI Engineering Conference Index* includes only individual papers presented at meetings, users must refer to *Engineering Index* for abstracts of journal articles and other primary literature. The computerized equivalent is *EI Meetings.*

27. **Electrical and Electronics Abstracts.** Vol. 1- . London: INSPEC, 1898- . monthly. ISSN 0036-8105. $1,000.00/yr. (Science Abstracts Series B.)

Probably the most important index of electrotechnology. It groups electrical and electronics engineering into general topics such as circuits and electronics, electron devices and materials, electromagnetics and communications, instrumentation and special applications, and power and industry. Complete bibliographic data is given for each entry. Abstracts more than 40,000 items from both primary and secondary literature. The monthly issues are organized into a subject classification system. The annual indexes are by author and subject. Cumulated author, subject indexes are available including the most recent covering 1981-1984. This cumulation may be purchased either separately or as a set. Price for the set is $1,455.00. This index is part of the INSPEC family of indexes which includes *Physics Abstracts* and *Computer and Control Abstracts* along with the *Current Papers* series of publications. The online equivalent is the database INSPEC.

28. **Electronics and Communications Abstracts Journal.** Vol. 1- . Riverdale, Md.: Cambridge Scientific Abstracts, 1967- . bimonthly. ISSN 0361-3313. $419.00/yr. (Formerly *Electronics Abstracts Journal*).

An international abstracting service covering electronic physics, electronic systems, electronic circuits, electronic devices, and communications. Approximately 5,000 items are abstracted each year from periodicals, government reports, conference proceedings, books, dissertations, and patents. Each issue includes author and subject indexes, which are cumulated annually. There is less emphasis on non US publications and greater emphasis on communications than in *Electrical and Electronics Abstracts.*

29. **Engineering Index Monthly and Author Index.** Vol. 1- . New York: Engineering Index, 1884- . monthly. ISSN 0162-3036. $1,290.00/yr. (Formerly *Engineering Index*).

This is the largest English language abstracting service in engineering. It indexes more than 2,050 journals, conferences, and miscellaneous serials published in 20 or more languages. Since much of engineering is applications oriented it is somewhat surprising that relevant patents and product literature are not indexed. The abstracts are arranged in classified subject order with each entry providing an abstract followed by the complete citation. The classified arrangement makes it somewhat difficult to retrieve information on emerging technologies or topics which fall into some of the larger subject breakdowns like "Flow of Fluids." A listing of these categories is available in *SHE: Subject Headings for Engineering*, which is available from Engineering Index.

A major change was made in EI in 1983 when only the subject of the entire conference or proceedings was abstracted in *Engineering Index*. Indexing for the individual papers is found in *EI Meetings* or *EI Engineering Meetings*. A list of journals along with their CODENs, known as PIE, "Publications Indexed for Engineering," is available as part of the subscription or as a separate purchase. The computerized equivalent is COMPENDEX.

30. **Index to IEEE Publications**. New York: IEEE Press, 1951- . annual. LC 75-646004. $200.00/yr. list, $100.00/yr. member.

This index provides access to articles in all journals, transactions, and magazines published by the IEEE as well as all published papers presented at over 140 IEEE-sponsored conferences during the year. Each volume includes summaries on IEEE periodicals, conference records and digests, special issues, standards and IEEE press books.

31. **Japanese Technical Information Service (JTA)**. Vol. 1- . Ann Arbor, Mich.: University Microfilms, 1985- . monthly. ISSN 0882-5246. $5,000.00/yr.

*JTA* is a cover-to-cover abstracting service which includes every article (whether referred or staff written), product announcements, and book reviews found in the covered Japanese journals. Each issue is indexed by author, authors affiliation, and subject. The monthly indexes cumulate into an annual index. Approximately 750 business and technical publications in all areas of engineering including electrical and electronic, and computers and information processing are covered with 150-word abstracts prepared by a U.S.-based staff of translators. Engineering, especially electrical, electronic, and computer engineering, makes up a major portion of the materials covered.

*Japanese Current Research (JCR)* is a companion publication comprised of copies of the table of contents of titles included in *JTA*. *JCR* is included in the subscription to *JTA*. This very expensive title is the only abstracting tool currently available which is specifically designed to answer the question "what are the Japanese doing in...?"

32. **Microelectronics Abstracts.** Vol. 1- . Evanston, Ill.: Lowry-Cocroft Abstracts, 1965- . semimonthly. $360.00/yr.

Abstracts of world literature in microelectronics are published on 5-by-8-inch coded, edge-notched index cards. This very limited subject index is designed to provide users with a quick way to maintain a literature file.

# 4   Current Awareness Services

The following periodicals are designed to provide rapid access by topic to current periodical and conference literature. Their goal is to overcome the time lag between publication, citation, and indexing of journals because in fast moving fields rapid access to new information can be very important. Scanning these current awareness tools on a regular basis will help keep readers up to date.

33.   **Current Contents: Engineering, Technology and Applied Sciences**. Vol. 1- . Philadelphia: Institute for Scientific Information, 1970- . weekly. ISSN 0095-7917. $245.00/yr. (Formerly *Current Contents/Engineering and Technology*, ISSN 0011-3395).

Consists of copies of the tables of contents from selected journals arranged by broad subject categories. Intended to serve the current awareness needs of engineers, each issue contains an author index and a key word index based on titles of articles. Authors' addresses are also listed. Regular features include an editorial, tables of contents from selected books and conferences, and a citation of the week. Each issue has a list of the journals which are included and a list of journals that have been added or dropped since the last issue. Copies of articles from the tables of contents are available from ISI.

34. **Japanese Technological Contents.** Tokyo: Nihon Faxon Co., 1985- . monthly. ISSN 0911-6516. $225.00/yr.

A compilation of tables of contents of Japanese academic journals in all areas of technology including electrical and electronic engineering. Each entry includes the publisher's name, English title, Romanized Japanese title, Japanese character title, ISSN number, frequency, price, text, and availability of English abstracts. A list of journals covered is provided for each issue. A key word title index is also provided. Translations services are available through the Faxon Tokyo office.

35. **Tech Notes: Computers.** Springfield, Va.: NTIS, 1981- . monthly. C51.17. $80.00/yr.

This publication and other *Tech Notes* are designed to inform a large audience of the technology developed by the U.S. government that is available for licensing as well as other types of development.

36. **Tech Notes: Electrotechnology.** Springfield, Va.: NTIS, 1981- . monthly. C51.17.

This current awareness publication from the Center for the Utilization of Federal Technology consists of short annotations of federally sponsored research which the center deems worthy of wider consideration in the area of electrical and electronics engineering and its applications.

## INSPEC Publications

37. **Current Papers in Electrical and Electronics Engineering.** Vol. 1- . London: INSPEC, 1969- . monthly. ISSN 0011-3778. $160.00/yr.

Published monthly to serve as the current awareness publication for *Electrical and Electronics Abstracts.* Coverage is international. Items are listed in a classified arrangement by broad subject with subdivisions. The entries have complete bibliographic data but do not include abstracts. There are no indexes for this publication.

38. **Current Papers on Computers and Control.** Vol. 1- . London: INSPEC, 1969- . monthly. ISSN 0011-3794. $160.00/yr.

The current awareness service of *Computers and Control Abstracts.* Organized in the same manner as *Current Papers in Electrical and Electronics Engineering.*

39. **Japan Update.** London: INSPEC, 1984- . monthly. $180.00/yr.

A new service from INSPEC designed to help users keep up to date with Japanese research in the areas of telecommunications; electronic devices and materials; instrumentation, computer hardware, and software; and computer applications and control technology. Articles are selected from approximately 249 Japanese language publications. Each citation is translated and printed on 4-by-6-inch cards which are mailed to subscribers. The intent is to get this information out to users before it is available online or published in the relevant abstract. Information is compiled weekly from the latest update to the INSPEC database.

INSPEC plans to substantially revise and enlarge their series of *Key Abstracts* in 1987. The new series will include three areas not previously covered: "Computer Communications and Storage," "Advanced Materials," and "Software Engineering." Other titles already in the series will be enlarged to include growth areas in electronics, computing, and physics. All of the INSPEC products are related; *Key Abstracts* and *Science Abstracts* are formed by adding abstracts to the bibliographic citations listed in *Current Papers*. The aim of *Science Abstracts* is to cover the entire field while the goal of the *Key Abstracts* is to allow researchers access to highly relevant information on very specific topics.

40. **Key Abstracts: Communication Technology.** Vol. 1- . London: INSPEC, 1975- . monthly. ISSN 0306-5588. $80.00/yr.
Indexes the fields of data, telegraphic, and radio communications; optical and satellite communications; image transmission; mobile radio; power communications; and switching systems. Arranged by broad subject classifications with an author index. There is no cumulative index.

41. **Key Abstracts: Electrical Measurements and Instrumentation**. Vol. 1- . London: INSPEC, 1976- . monthly. ISSN 0307-7977. $80.00/yr.
Arranged like the other *Key Abstracts*, this index covers the fields of measurement science, measurement standards, measurement equipment, instrumentation systems, etc. Has no cumulative index.

42. **Key Abstracts: Electronic Circuits**. Vol. 1- . London: INSPEC, 1975- . monthly. ISSN 0306-557x. $80.00/yr.
Power electronics, amplifiers, oscillators, modulators, demodulators, mixers, pulse and digital circuits, logic and memory circuits, microprocessors, microcomputers, etc. are indexed. No cumulative index.

43. **Key Abstracts: Industrial Power and Control Systems.** Vol. 1- . London: INSPEC, 1975- . monthly. ISSN 0306-5596. $80.00/yr.
Treats the fields of electric drives, electric heating, air conditioning, refrigeration, materials handling, control, and electric power applications in metallurgical, manufacturing, chemical, oil, and textile as well as other industries. No cumulative index.

44. **Key Abstracts: Physical Measurement and Instrumentation.** Vol. 1- . London: INSPEC, 1976- . monthly. ISSN 0307-7969. $80.00/yr.
Subjects covered include measurement and metrology; mechanical measurements and techniques; thermal instruments and techniques; pressure measurements and techniques; and optical, magnetic, and electrical measurements and techniques. Monthly author index, but no annual cumulation.

45. **Key Abstracts: Power Transmission and Distribution.** Vol. 1- . London: INSPEC, 1975- . monthly. ISSN 0306-5561. $80.00/yr.
The fields of AC and DC transmission, distribution networks, lines and cables, connectors, transformers, switchgear, power conversion, and system protection and measurement are covered. No cumulative index.

46.   **Key Abstracts: Solid-State Devices.** Vol. 1- . London: INSPEC, 1975- . monthly. ISSN 306-5537. $80.00/yr.

Includes coverage of the fields of semiconductor junctions and interfaces, diodes, field effect devices, charge-coupled devices, thyristors, ICs, photoelectric devices, solar cells, lasers, light emitting devices, semiconductor logic, and storage devices.

47.   **Key Abstracts: Systems Theory.** Vol. 1- . London: INSPEC, 1975- . monthly. ISSN 0306-5553. $80.00/yr.

Among the areas dealt with are adaptive systems, optimal control, stability, simulation, discrete-time systems, artificial intelligence, pattern recognition, information theory, and man-machine systems.

# 5 Proceedings and Other Speciality Indexes

48. **Directory of Published Proceedings.** Vol. 1- . Harris, N.Y.: InterDoc, 1965- . 10 times/yr. ISSN 0012-3293. $225.00/yr. (Semt: Science/Engineering/Medicine/ Technology).

Commonly known as "InterDoc," this is bibliographic directory of preprints and published proceedings of congresses, conferences, symposia, meetings, etc., which have been held worldwide. Includes editor, location, and subject/sponsor indexes as well as an annual, cumulated index. The entries are arranged in chronological order. This arrangement by date and location is especially useful because many users remember only the date and the city for conferences; with this index it is possible to quickly find the title or sponsor of a conference and more importantly to see if there were published proceedings. Ordering information is provided for all entries. This is a companion to *MInd* (entry 52).

49. **Government Reports Announcements and Index.** Vol. 74- . Springfield, Va.: NTIS, 1974- . 25 times/yr. ISSN 0097-9007. PB85-900116. $325.00/yr., including annual index cumulations. (Formerly *Bibliography of Scientific and Industrial Reports,* U.S. Government Research Reports, and *Government Reports Announcements*).

Indexes of U.S. Government-sponsored research and development publications prepared by universities, government laboratories, consultants, and contractors. The biweekly indexes include bibliographic citation, abstract and key word, author, corporate author, and contract and report number indexes. These indexes are then cumulated annually. In addition to serving as an index, *Government Reports Announcements and Index* can also be used as an acquisitions tool as each entry lists whether the document may be purchased from NTIS and at what price. The abstract section is divided into subjects. Section 9, "Electronics and Electrical Engineering," is the most directly related to electrotechnology. Available on DIALOG and BRS.

50.  **Index to Scientific and Technical Proceedings (ISTP).** Philadelphia: Institute for Scientific Information, 1978- . monthly. ISSN 0149-8088. $650.00.

An index to published proceedings that have appeared in almost all forms including books, reports, preprints, or as part of a journal. Coverage is limited to those proceedings in which the majority of the material is being printed for the first time. Publications consisting only of abstracts or digests are not included. There are indexes by conference sponsor, author/editor, meeting location, permuted subject, and annual cumulations. A unique feature in *ISTP* is access to published proceedings which have appeared in journals.

51.  **International Aerospace Abstracts.** Vol. 1- . New York: AIAA/TIS, 1960- . semimonthly. ISSN 0020-5842. $700.00/yr. plus $400.00/yr. for annual cumulative index.

The world's published literature in the areas of aeronautics, space science, and technology is covered. Publications treated include periodicals, books, meetings, and conference proceedings issued by professional societies and academic organizations. *International Aerospace Abstracts* is published in coordination with *STAR* (see entry 53). The abstracts are organized by sections, with the following sections the most relevant to electrotechnology: section 32, "Communications;" Section 33, "Electronics and Electrical Engineering;" section 60, "Computer Operations and Hardware;" and section 76, "Solid-State Physics." Indexes which include subject, author, contract and report number, and accession number, appear in each issue and are cumulated annually. All referenced items are available from the AIAA Technical Information Service. Available on DIALOG, Mead, and NASA RECON.

52.  **MInd: The Meetings Index.** Vol. 1- . Harrison, N.Y.: InterDoc, 1984- . bimonthly. ISSN 0739-5914. $325.00/yr. (Semt: Science/Engineering/ Medicine/ Technology).

*MInd* is a companion publication to the *Directory of Published Proceedings*. The purpose of this index is to classify and index forthcoming meetings, conferences, courses, symposia, seminars, and summer schools on a worldwide basis. All entries, therefore, pertain to future meetings. Indexes include: key word, sponsor, location, date, and contact person.

53.  **Scientific and Technical Aerospace Reports (STAR).** Vol. 1- . Airport, Md.: NASA/TIF, 1963- . 25 times/yr. ISSN 0036-8741. $85.00/yr. including annual cumulative indexes.

*STAR* is a major component of the comprehensive indexing system of the NASA Information System which covers aeronautics, space, and the supporting disciplines. Unlike *International Aerospace Abstracts (IAA)*, which indexes periodicals and conferences, *STAR* indexes reports issued by U.S. government agencies, domestic and foreign institutions, universities, and private firms that have done research under contract to NASA. Along with this coverage of report literature, *STAR* also covers NASA-owned patents and NASA-sponsored dissertations and theses. Like *IAA*, this index is organized into sections of which section 32, "Communications;" section 33, "Electronics and Electrical Engineering;" section 60, "Computer Operations and Hardware;" and section 76, "Solid-State Physics" are of particular interest. There are indexes by subject, author, corporate source, and contract and report number. Available on DIALOG, Mead, and NASA RECON.

54.   **World Meetings: Outside United States and Canada.** Vol. 1- . New York: Macmillan Publishing, 1968- . quarterly. ISSN 0043-8677. $135.00/yr.

Consists of a two-year listing of future scientific and technical meetings which will be held outside of the United States and Canada. A companion publication to *World Meetings: United States and Canada* (entry 55). Key word, location, date, sponsor, and deadline indexes are included. Each entry has sponsor, general information, attendance restrictions, location, and exhibition and publication information, if known.

55.   **World Meetings: United States and Canada.** Vol. 1- . New York: Macmillan Publishing, 1963- . quarterly. ISSN 0043-8693. $125.00/yr.

A listing of scientific and technical meetings that will be held within the next two years in the United States and/or Canada. Organized in the same manner as its companion publication, *World Meetings: Outside United States and Canada.*

# 6  Databases

A major development over the last decade in information retrieval has been the increasing availability of machine-readable bibliographic databases. These databases are now commonly available in research laboratories; public, industrial, academic, and government libraries; and publishing houses. One of the first bibliographic databases was NASA RECON, which was the forerunner of the current DIALOG system. It was designed as a research project for NASA by the Lockheed Corp. when NASA needed a way to index and organize the vast amount of information being collected as a result of work on the Apollo moon program.

This basic research and the concurrent development of the new computer-controlled printing presses laid the groundwork for the new information retrieval industry. It became evident that the computer tapes being developed to print the paper indexes could be made searchable with the addition of appropriate software using a system similar that developed by Lockheed for NASA. Several companies began to develop vendor-based searching: Lockheed's DIALOG, SDC's ORBIT, and finally BRS. These vendor services took the tapes from publishers, added appropriate search capabilities, and then sold the new service to research centers, libraries, government agencies, etc. While most literature searching is still associated with the large vendors, a number of databases are beginning to be marketed directly to the consumer, such as WILSONLINE, which is being sold directly to both individuals and libraries.

The systems used by trained searchers as intermediaries for the researcher are command driven. To use such a system, the searcher, through training and experience, must learn the commands needed to activate the computer and retrieve relevant information, usually in the form of bibliographic citations. Because these systems were not designed for casual or infrequent use, over the last several years menu driven, simplified searching systems have been developed, which can be used without extensive training, practice, or manual reading. They were designed to be used by people who do not search on a regular basis. The two best known are BRS AFTER DARK and DIALOG's KNOWLEDGE INDEX. Both are aimed at users with access to microcomputers and modems.

Bibliographic databases have a number of advantages over paper abstracts: they can handle large amounts of bibliographic data simultaneously, the data can be manipulated quickly, and most importantly, words or terms can be intersected to create tailored results. Therefore, users no longer have to rely on controlled vocabulary lists, cross-references, or their own ideas about the subject of articles because it is possible to search any word or phrase that might appear in an abstract or title. Also it is now possible to search for very specific articles, such as expert systems applied to circuit design or protective relays for high voltage lines.

Until recently most commercial development was in the area of computerized bibliographic databases including the group now known as Library Utilities—OCLC, RLIN, and WLN. The area of nonbibliographic database production was primarily the function of the individual researcher or research laboratory and these databases were available only by word of mouth. This is now changing and many users believe that the next big growth area for searching will be in nonbibliographic databases.

While the following section is primarily concerned with bibliographic databases, it does include several nonbibliographic databases. Because new databases and services, such as APPLIED SCIENCE AND TECHNOLOGY INDEX on CD-ROM, are coming online regularly, users should consult with a librarian or vendor concerning current availability and costs.

56. **AEROSPACE DATABASE.** New York: American Institute of Aeronautics and Astronautics. 1962- .

Coverage extends from 1962 to the present for materials from STAR and from 1963 for materials from IAA. Abstracts are included beginning with 1972. Areas covered include worldwide materials of scientific and technical literature with an emphasis on aeronautics, astronautics, celestial mechanics, avionics (electronics applied to aerospace), and space. This database corresponds to NASA RECON, which in the past was available only to NASA contractors and includes classified materials. These classified materials have been dropped from the commercial system and at the same time searching has been made easier. One major change is that use of the NASA THESAURUS is no longer imperative.

57. **CASSIS.** Washington, D.C.: U.S. Department of Commerce, Patent and Trademark Office, 1879- .

This database is available free of charge in any Patent Depository Library, a list of which is printed in the front of the first *Official Gazette* published each month. Using this database it is possible to search the *Manual of Classification*, the *Index of Patents*, and all patent abstracts since 1982. A limited retrospective input of more abstracts is planned. It is not possible to search by inventor, assignee, or title of patents. CASSIS is most effective for retrieving all patents in a particular class and subclass, making it no longer necessary to search the *Index of Patents* year by year, becoming entangled in the web of changing classifications and subclasses.

58. **CLAIMS/U.S. PATENT ABSTRACTS.** Alexandria, Va.: IFI/Plenum Data Company, 1950- .

Covers patents listed in the weekly issues of the United States Patent and Trademark Office, *Official Gazette*. The early years of the database, 1950-1963, cover only patents from the "Chemical" section of the *Official Gazette*. In 1963, bibliographic information on patents from the "Mechanical" and "Electrical" sections were added. Another major change to the database was made in 1971 when abstracts were added for all patents.

59. **COMPENDEX. Engineering Index Online Equivalent.** New York: Engineering Index Information, Inc., 1970- .

COMPENDEX is one of the more heavily used databases on both the BRS and DIALOG systems. Since 1983 a number of major changes have been made in the databases. Perhaps one of the most important is the addition of a language code for English so that it is now possible to retrieve only articles which are in English. Another recent addition is the inclusion of treatment codes, which have proven to be very useful in the INSPEC database and should be just as useful in COMPENDEX. Users are regularly looking for practical or experimental information and this has historically been very difficult to find without treatment codes. Two other interesting searchable fields are class codes and CODEN. Class codes are a quick way of getting all available material on a topic whether that topic is specifically used as an index term or found in the title or abstract. A list of class codes is available from the publisher.

60. **DERWENT WORLD PATENTS INDEX.** London: Derwent Publishers, 1968- .

Divided into searchable files called WPI (World Patents Index) and WPIL (World Patents Index Latest), this index contains nearly three million inventions represented by six million patents from 30 patent issuing countries. Since 1981 all citations include an abstract and title written by the editors at Derwent in an attempt to overcome the sometimes vague titles given by inventors. For example, the title for the Frisbee patent is "A Levitating Disk." Neither the abstract nor the title mentions the words flying or saucer or even toy. The rewritten Derwent abstract and title correct this problem.

One of the most important features of WPI is the capability of linking patents relating to the same invention from different countries into a single patent "family." As a result each related or equivalent patent is grouped with first patent or patent application input by Derwent to form a family. With a German application number a user could see if a patent protection was also granted in Great Britain, the United States, or Japan. Subscribers to Derwent paper and microfiche products receive a reduced searching rate.

61. **EI ENGINEERING MEETINGS.** New York: Engineering Information, 1982.

This online database is a companion to COMPENDEX which indexes individual papers from published meetings, symposia, and conferences. Since 1983, COMPENDEX has indexed conferences only by the main entry or subject of the entire conference. Each entry includes the number of papers and the EI Meetings accession number, which can be used to retrieve the titles of the individual papers. Items entered before 1984 do not include abstracts. The paper equivalent is *EI Engineering Conference Index*.

62. **ELECTRIC POWER DATABASE.** Palo Alto, Calif.: Electric Power Research Institute, 1972- .

Includes references to research and development projects of interest to the electrical power industry. It covers U.S. and Canadian research on issues related to fossil fuels, nuclear power, transmission, economics, and environmental assessment. The records include abstracts for both past and ongoing projects. Although coverage is primarily of projects funded by EPRI, some other research of interest is covered. This database corresponds to the printed *Digest of Research in the Electric Utility Industry*. EPRI funds a considerable amount of research and this is a good place to identify projects they have funded in the past.

63. **ELECTRONIC PROPERTIES INFORMATION CENTER (EPIC).** West Lafayette, Ind.: CINDAS, 1966- .

Online database corresponding to *Electronic Properties of Materials: A Guide to the Literature*. Produced by ThermoPhysical and Electronic Properties Information Analysis Center, this nonbibliographic database provides the actual data along with a reference to where this data was previously published. Many indexing sources like *Chemical Abstracts, Engineering Index*, and the INSPEC do not index or reference data that is presented in the literature. CINDAS does. It is therefore possible to retrieve dielectric constants and experimental data that has been in the published literature. Available as tapes from EPIC.

64. **ELECTRONICS & COMPUTERS (ELCOM).** Bethesda, Md.: Cambridge Scientific Abstracts, 1977- .

Corresponds to *Electronics & Communications Abstracts Journal* and *Computer & Information: Systems Abstracts Journal*. Covers electronic physics, electronic systems and applications, electronic circuits, electronic devices, communications, computer software, computer applications, computer mathematics, and computer electronics. Abstracts are taken from international sources including periodicals, conference proceedings, books, dissertations, government reports, and patents.

65. **EMIS (Electronic Materials Information Service).** London: Institution of Electrical Engineers, n.d. (Marketed by General Electric Co.).

This primarily numeric database provides information on the properties of materials used in the design of solid-state circuits and other electronic devices. It also has a supplier index so that users can determine who manufactures a material with the specific characteristics and properties needed. The supplier index gives the address, phone numbers, and salesman's name. Recently EMIS added a new feature making it possible to input semiconductor data directly into a subset of the database.

66. **INDUSTRY AND INTERNATIONAL STANDARDS DATABASE.** Englewood, Colo.: Information Handling Services, n.d.

Provides access to active and historical documents for federal specifications, federal standards, joint army-navy specifications, military specifications, and military standards. It also provides access to U.S. voluntary standards, as well as standards from Germany (DIN), the United Kingdom, and others. This database makes it possible to see if anyone has a standard on a particular topic. Full ordering information is given as well as cartridge and frame citations to Information Handling Services (IHS) subscription services for standards and specifications. Available on BRS or directly from vendor. Standards may be ordered online.

67. **INPADOC.** Vienna: International Patent Documentation Centre, 1968- .

Includes patents issued since 1968 from 51 national and regional patent offices. Searchable by inventor, assignee, language equivalent, and subject of the invention. A competitor to Derwent.

68. **INSPEC.** London: IEE, 1967- .

The online equivalent of three sections of *Science Abstracts*—"Computers and Control," "Electrical and Electronics Abstracts," and "Physics Abstracts." May be searched by each section separately or all at the same time. One of the strengths of the INSPEC database is the inclusion of treatment codes for each abstract, making it possible to search for concepts which are not normally used for indexing. For example, how does the searcher distinguish between theoretical and experimental papers. To meet this need, since 1971 INSPEC has assigned treatment codes to each paper. These codes allow users to search for papers with the following concepts: applications, bibliography/literature survey, economic, general or review, new developments, practical, theoretical, and experimental. In 1985 a new treatment code was added to cover product reviews. INSPEC is a good source for commercial and technocommercial articles which are often difficult to retrieve in other online systems and even more difficult to retrieve manually.

69. **JAPAN INFORMATION CENTER OF SCIENCE AND TECHNOLOGY: ELECTRONICS AND ELECTRICAL ENGINEERING.** (JICST File: Electronics and Electrical Engineering). Tokyo: Nippon Kagaku Gijutsu Joho Senta, 1970- .

A companion to *Current Bibliography on Science and Technology: Electronics and Electrical Engineers*, this database covers electrical engineering, electronics, control technology, communications, and computer technology. A thesaurus is available. Tapes may be purchased from the producer.

70. **JAPIO: JAPANESE PATENT INFORMATION ORGANIZATION.** Tokyo: Japio, 1976- .

Provides access to over one million Japanese patents in English. Each record provides patent number, priority data, inventor, international classification number and the title, and abstract in English. All areas are detailed, including electronics, communications, and instrumentation. The database is updated monthly. Other features include online document ordering, cross file searching with Derwent's WORLD PATENTS INDEX, and patent cluster searching with APIPAT. Available on SDC.

71. **MENU/INTERNATIONAL SOFTWARE DATABASE.** Fort Collins, Colo.: International Software Database Corp., n.d.

A comprehensive listing of commercially available software packages. A paper version is available from Elsevier.

72. **NBS ALLOY DATA BASE.** Washington, D.C.: U.S. Department of Commerce, National Bureau of Standards, 1979-1981.

Known by the name ALLOY, this database covers the physical, mechanical, electrical, magnetic, and thermodynamic properties of metals and alloys. The tapes are available from NBS.

73. **PATSEARCH.** New York: Pergamon InfoLine, Inc.: 1971- .

Contains bibliographic descriptions and abstracts for all U.S. patents issued since 1971. Each patent entry has both U.S. and European classification codes, as well as priority filing information and references from the first page of the patent. These references are relevant to patent and nonpatent literature. Available on INFOLINE and as Video PATSEARCH. Video PATSEARCH incorporates a display of the graphic from the front page of the patent, allowing the user to see a drawing or chemical formula of the patented device.

74. **SOFTWARE ENGINEERING BIBLIOGRAPHIC DATABASE.** Rome: IIT Research Institute, 1968- .

Covers software engineering, software reliability, software quality, software tools, and modern programming practices. Most of the references come from journals, government reports, and proceedings. Tapes are available from producer.

75. **STANDARDS AND SPECIFICATIONS.** Bethesda, Md.: National Standards Association, 1950- .

Provides access to all active government and industry standards, specifications, and related documents with nearly 104,000 bibliographic records. Each entry includes issuing organization, and, where relevant, Federal Supply Classification Code, cancelled or superseded status, and information on whether the standard has been adopted by a U.S. government agency or designated an American National Standards Institute (ANSI). This is the online equivalent of NBS SP 329.

76. **STDS.** Englewood, Colo.: Information Handling Service, n.d.

Consists of voluntary engineering standards from all standardizing bodies in the United States and selected foreign national and international bodies. Citations come from IHS databases and from the NBS Voluntary Engineering Standards database. Each citation includes the location with the IHS subscription files for the original document. Available on BRS.

77. **VIDEOLOG.** Santa Clara, Calif.: Videolog Communications, 1985- .

A videotext catalog of electronic product data consisting of the *Harris Directory: Who's Who in Electronics, D.A.T.A. Books,* and product information and specifications provided by manufacturers. The database consists of 18,000 product categories, 14,000 manufacturers, up to 15 parameters on 1,000 distinct semiconductor device types, and over 500,000 types of integrated circuits, microcomputer systems, and discrete semiconductors. Uses NAPLPS (North American Presentation Level Protocol Syntax), modem and an IBM-compatable personal computer with color monitor and graphics board. Data are updated quarterly. Available through THE SOURCE or via a gatekeeper to TELENET or TYMNET.

# 7 Computer Software

The growing importance and availability of computers has had an effect on all aspects of engineering from pure research and teaching to the everyday work environment. Electrical and electronics engineering has been more affected by the computer revolution than some other engineering areas, primarily because much of the development of computers has been a direct result of engineers' work.

The growing availability of computers has also had a dramatic impact on the development of software. Until quite recently engineers had to write, or have written for them, all the software that was needed. It is now possible to purchase generic software for use in many engineering applications as well as software for specific engineering problems. For example, several recent articles have been written about using a spread sheet program to solve electrical engineering calculations.

With this growth in software, the need for access to commercially produced software has arisen. The following section brings together materials that provide access to available software. A selected number of books which reproduce computer programs and codes are also included.

78. **Bowker's Microcomputer Software in Print.** New York: R. R. Bowker, 1985. 2v. plus supplement. ISBN 0-8352-1944-5. $95.00/2v.; $44.50/suppl.

Volume 1 contains "Software Products Profiles," which gives an annotated listing of available microcomputer software. The listings are divided into 100 categories with over 2,500 subcategories. Volume 2 includes the "Product Name Index," a listing by brand or trade name, the "Computer System Index," a listing of programs by computer, and a "Vendor Index," which gives the names, addresses, telephone numbers, dates founded, number of employees, sales, subsidiary relationship, primary customers, description of the product line, and a profile of the company.

79. **Circuits and Software for Electronic Engineers.** Edited by Howard Bierman. New York: McGraw-Hill, 1983. 344p. illus. no index. LC 83-9806. ISBN 0-07-005243-3. $35.00.

A listing of software published by *Electronics Magazine* between 1980-1982 for specific design problems. Some of the areas covered are converters, encoder/decoders, instrument circuits, and software/computer. Specific software examples include a code for the TI-59 program that tracks satellites in elliptical orbits and an 8-bit DMA controller that handles 16-bit data transfers. The lack of an index limits the usefulness of this book.

80. **Computer Programs for Electronic Analysis and Design.** By Dimitri Bugnolo. Englewood Cliffs, N.J.: Prentice-Hall, 1983. 288p. index. LC 82-23167. ISBN 0-8359-0874-7. $21.95.

A collection of programs or codes for analysis and design of solid-state electronic circuits, written primarily for micro-computers and hand-held calculators. This book is popular with students.

81. **Computer-Readable Databases: A Directory and Data Sourcebook.** Edited by Martha E. Williams. Chicago: American Libraries Association. 1985. 2v. index. ISBN 0-8389-0415-7. $157.50.

First published in 1976, this directory covers business, law, medicine, social sciences, humanities, technology, engineering, and consumer information. Coverage is international and makes available for the first time both word-oriented and numeric databases. Approximately 2,000 databases are covered. Access is through subject, producer, online vendor, and name indexes. Each entry includes name, previous name, update frequency, producer, subject matter, data elements, user aids, and availability. It is also sold as separate volumes: *Science, Technology and Medicine* (ISBN 0-8389-0416-5), and *Business, Law, Humanities, Social Sciences* (ISBN 0-8389-9417-3).

82. **Directory of Computer Software.** Springfield, Va.: NTIS, 1985. 271p. index. PB 85-162121. C 51.11/2. $17.50.

Abstracts of 1,300 computer programs compiled by 100 federal agencies and/or their contractors are listed. The programs are indexed by agency, accession number, subject, hardware, and language.

83. **Directory of Computer Software Applications: Electrical and Electronics Engineering.** Springfield, Va.: NTIS, 1978. PB 284-924. $17.50/paperback.

Some 540 abstracts of software applications in the areas of antenna design, computerized simulation, microwave oscillators, and semiconductor devices are covered. The text is arranged by subject with each entry giving the NTIS order number, so the programs/code may be ordered from NTIS. A supplement was issued in 1983.

84. **Directory of Computerized Data Files.** Springfield, Va.: NTIS, 1985. 339p. index. PB 85-155174. C 51.11/2-2. $17.50.

The purpose of this volume is to provide current information on the availability and content of federal machine-readable data files. More than 100 data files are described.

85. **Engineering Microsoftware Review.** Vol. 1- . Medford, N.J.: Learned Information, 1985- . monthly. $48.00/yr.

Reviews are based on the findings of a panel of industrial and academic engineers who examine the performance of the package in solving problems, ease of use, and computers on which the packages will run. Recent issues have included reviews of statistical packages, linear programmings, chart and curve graphics, and project planning.

86. **ESE Engineering Software Exchange.** Vol. 1- . Yonkers, N.Y.: Elsevier Science Publishing, 1983- . bimonthly. ISSN 0743-2984. $48.00/yr.

Designed to provide the latest engineering software applications for micro and mini systems. Critical reviews, comparative reviews, and new developments are regular features. Readers may list software for sale or trade in a special section.

87. **Handbook of Electronic Design and Analysis Procedure Using Programmable Calculators.** By Bruce K. Murdock. New York: Van Nostrand Reinhold, 1979. 525p. index. LC 79-15122. ISBN 0-442-26137-3. $36.50. (VNR Electrical/Computer Science and Engineering Series).

Contents include 26 programs pertaining to network analysis, 15 to filter design, nine to electromagnetic component design, five to high-frequency circuit design, and four to engineering mathematics. Includes a list of abbreviations and a bibliography. Although the programs are tailored to the Hewlett-Packard fully programmable calculators, the annotated program flowcharts can be used as the basis for generating programs in other languages for other calculators.

88. **Hewlett-Packard Software Catalog: Summer 1984.** New York: John Wiley, 1984. 384p. index. ISBN 0-471-81912-3. $25.00.

Divided into sections by major applications and by vendor within each section, this catalog lists software available for the Hewlett-Packard 150 calculator. Each summary contains a complete description, the required configuation, support, training information, and where to order. Indexed by software names and vendors.

89. **IBM PC Programs in Science and Engineering.** By Jules H. Gilder. New York: Hayden Book Co., 1984. 256p. index. ISBN 0-81949761-2. $18.95.

Contains a software library of approximately 100 programs. Of particular interest are programs which solve problems in basic electricity and electronics, and computer-aided designs of amplifiers, and active and passive filters.

90. **Introduction to Operating Systems.** 1st rev. ed. By Harvey M. Deitel. New York: Addison-Wesley, 1984. 704p. illus. index. LC 83-7153. ISBN 0-201-14501-4. $39.95.

Describes the VAX, UNIX, CP/M, MVS, and VM operating systems. Part 1 discusses the history of operating systems, hardware, software, and firmware, while part 2 covers process management including asynchronous, concurrent, and deadlock. Storage management and disk scheduling are detailed in parts 3, 4, and 5. Parts 6 and 7 cover performance measurements and networks, while part 8 consists of case studies. References and explanations of terminology are found at the end of each section.

91. **Science and Engineering Programs for the IBM-PC.** By Cass R. Lewart. Englewood Cliffs, N.J.: Prentice-Hall, 1984. 240p. index. LC 83-15955. ISBN 0-13-794934-0. $39.95.

Code for 20 programs in electronic engineering, number theory, computer program design, and data communications are covered.

92.   **Software Abstracts for Engineers (SAFE).** Vol. 1- . New York: CITIS, 1984- . quarterly. ISSN 0790-150x. $140.00/yr.

Aims to provide working engineers with the details of the latest computer programs. Each listing includes the address of firm producing the software, names of the computers supported, types of peripherals required, services available from the vendor, price, and a summary of the program. Subject access to software is through a key word index. An annual index is published. Of special interest are the lists of books and periodicals which review programs, cover programming languages or techniques, and discuss applications of computer systems or graphics packages.

93.   **Software Catalog: Science and Engineering.** 3rd ed. New York: Elsevier Science Publishing, 1986- . 612p. index. LC 84-6154. ISBN 0-444-00925-6. $49.50.

With coverage of more than 4,300 existing micro and mini computer packages, this is a reference to information on availability, price, application, and compatibility. The index allows access by vendor, computer system, operating system, programming language, microprocessor, application, and key word. The software is arranged in the following categories: industrial, scientific, professional, and system/utilities. This is the paper equivalent of the online database MENU/INTERNATIONAL SOFTWARE DATABASE on DIALOG.

94.   **Softwhere.** Minneapolis, Minn.: Moore Data Management Service, 1984. ISBN 0-918451-30-2. $35.00.

This comprehensive directory of engineering software allows easy comparison of functions, prices, and compatible systems. Arrangement is by broad subject. The listing of available software by computer system is of special interest.

# 8 Book Review Sources

Not only are there no sources specifically designed to review books in electrical and electronic engineering, there are no review sources aimed only at engineering. Therefore the following review sources cover all of science and technology, but include separate sections on electrical engineering. Review publications are not the only place to find reviews. Many magazines, for example, run a regular book review column. However, the easiest way to keep on top of new books is to watch for publishers' advertisements in electrical and electronics trade publications.

95.  **Aslib Book List.** London: Aslib, 1935- . monthly. ISSN 0001-2521. $38.00/yr.
Arranged by Universal Decimal Classification, with a broad subject index, this publication leans heavily toward books published in the United Kingdom. Areas covered include science, technology, medicine, and the social sciences. Each review is written by a subject specialist and given an audience rating code. For example "C" is the designation for a book for an advanced audience.

96.  **New Technical Books: A Selective List with Descriptive Annotations.** New York: Research Libraries, The New York Public Library, 1915- . monthly except August. ISSN 0028-6869. $15.00/yr.

The reviews are arranged by Dewey Classification. Each issue has separate author and key word title indexes, which are cumulated in December. The reviews are reasonably timely and aimed at the large public and academic library market. Each review includes a suggested audience or type of library. The low subscription fee makes this an affordable title for most libraries.

97.   **Technical Book Review Index.** Pittsburgh, Pa.: JAAD, 1977- . monthly except July and August. ISSN 0040-0890. $35.00/yr. (Continuation of *Technical Book Review Index* published by Special Libraries Association 1934-1976).

Provides citations and quotations taken from technical book reviews which were originally published in some 3,500 scientific technical and medical periodicals. Reviews run generally 13-15 months after publication of the title reviewed.

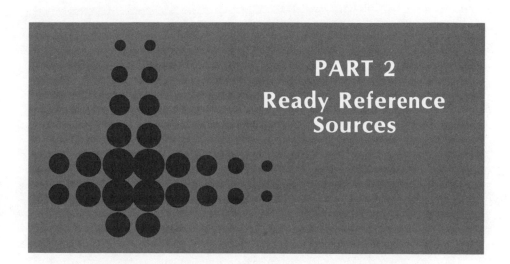

# PART 2
## Ready Reference Sources

# 9  Encyclopedias

Encyclopedias are a basic reference tool in any field. Electrotechnology, which is no exception, is made up of ideas and research from many areas in science and technology including physics, acoustics, materials science, and, of course, electrical and electronic engineering. This complexity and overlapping necessitates the use of scientific encyclopedias.

A good example of this overlapping of areas is microwaves, which is a field in and of itself. However, it is strongly related to signal processing, which is a subfield of electrical engineering. Many communications systems are based on sending and receiving microwave signals both overland and from satellites. For engineers and students who are unfamiliar with a related technology like microwave engineering, encyclopedias play an important role.

A useful encyclopedia gives good coverage of the field in question and the related areas. Two kind of entries are typically found, those defining the topics and those consisting of overview articles. The definitions and articles should consist of the most up-to-date information available. The articles should be written in a clear style without the use of undefined, specialized vocabulary and acronyms. Another characteristics of a good encyclopedia is articles written by named authorities with selected references. It is generally assumed that all books are written by authorities or experts. However, in the case of reference books it is beneficial for users to be able to judge the credentials of the authors for themselves. This is possible when the authors of each section are named.

98.   **Advances in Electronics and Electron Physics.** New York: Academic Press, 1948-. 2 times/yr. plus irregular supplement. ISSN 0065-2539. $164.00/yr.

Subjects include electronics, electron physics, electron microscopy, and the related areas of digital image processing and hybrid electronics. The signed articles are, essentially, critical reviews of current work. Each volume has author, subject, and cited author indexes. This series should be owned by any library with a serious electronic or electronic physics collection.

99.   **Encyclopedia of Computer Science and Engineering.** 2nd ed. Edited by Anthony Ralston and Edwin D. Reilly, Jr. New York: Van Nostrand Reinhold, 1983. 1664p. illus. index. LC 82-2700. ISBN 0-442-24497-7. $87.50.

This heavily revised second edition contains 550 articles, each with references and suggested reading lists. The encyclopedia covers all aspects of computer science and computer engineering, from electronic funds transfer to procedure oriented languages. The appendixes are extensive and cover abbreviations, acronyms, and numerical tables. Also includes compilations of computer science and engineering research journals, names of universities offering PhDs in computer science, lists and definitions of high-level languages, and a glossary of basic terms in five languages. Of special interest is the information on software, the mathematics of computing, and the theory of computation. This useful up-to-date, one-volume encyclopedia should be in all types of collections, from public to personal libraries.

100.   **Encyclopedia of Computer Science and Technology.** Edited by Jack Belzer. New York: Marcel Dekker, 1975-1981. 16v. LC 74-29436. ISBN 0-8247-2251-5. $99.75/ea.

A monumental 16-volume encyclopedia with a separate index, which provides access by both subject and author. Computer technology as well as fields in which computer technology has widespread usage are covered. Includes over 2,000 articles on 10,000 pages. Aimed at the needs of hardware specialists, programmers, systems analysts, engineers, and operations researchers, the treatment is scholarly and exhaustive, yet suitable for the nonspecialist. All libraries with more than a casual interest in electronics or computer science should own this set.

101.   **Encyclopedia of Computers and Electronics.** Chicago: Rand McNally, 1983. 140p. illus. index. LC 83-61911. ISBN 0-528-82390-2. $65.00.

Developed as a guide to electronics and computers, this book provides a cursory look at these fields. Brief explanations of microwaves, binary numbers, computer graphics, and other areas are given. A multitude of color illustrations supplements the text. Subjects appear to be listed in chronological order, but this organization is not actually spelled out. A glossary and index round out the volume. This work could serve as a supplemental source in school media centers and public libraries. However, not enough information is provided for this publication to be a primary source for a beginner.

102.   **Encyclopedia of Physical Science and Technology.** Edited by Robert A. Meyers. New York: Academic Press, 1987. 15v. illus. index. LC 86-1118. ISBN 0-12-226901-2. $1,975.00/set.

Aimed specifically at the physical sciences community, this encyclopedia consists of long review-type articles with an emphasis on mathematics. For example, calculus is used in many of the articles. The set consists of 550 articles with 4,000 titles listed in the bibliographies.

Volume 15 serves as an index, which, due to the length and inclusiveness of the articles, must be consulted. Each article includes the author's name, a short outline, a glossary of terms, and a bibliography. The coverage of electrical and electronics engineering is very strong. Other areas of interest given in-depth coverage include the properties of materials, and basic theories and mathematics used in the physical sciences.

103.   **Encyclopedia of Semiconductor Technology.** Edited by Martin Grayson. New York: John Wiley, 1984. 941p. illus. index. LC 83-21587. ISBN 0-471-88102-3. $99.95. (Encyclopedia Reprint Series).

An encyclopedia consisting of articles, graphs, drawings, figures, and reference materials taken in their entirety from original volumes of the *Kirk-Othmer Encyclopedia of Chemical Technology*. Nearly all aspects of semiconductors are covered, including methods of manufacture, properties, and use of common elements. The entries are arranged in alphabetical order, with extensive cross-references and references. Coverage is diverse, ranging from amorphous magnetic materials and digital displays to light emitting diodes and semiconductor lasers. This book should be purchased only by libraries that do not own a complete Kirk-Othmer  or have access to its online equivalent.

104.   **McGraw-Hill CD-ROM Science and Technical Reference Set.** New York: McGraw-Hill, 1987. CD-ROM. ISBN 0-07-852247-1. $300.00.

This compact disc contains 7,300 articles from the *McGraw-Hill Concise Encyclopedia of Science and Technology* along with 98,000 terms from *McGraw-Hill Dictionary of Scientific and Technical Terms*. It is designed to run on the IBM AT or XT computer with 640K RAM. The software provides access by key title or phrase.

105.   **McGraw-Hill Encyclopedia of Electronics and Computers.** Edited by Sybil P. Parker. New York: McGraw-Hill, 1984. 984p. illus. index. LC 83-9897. ISBN 0-07-045487-6. $72.50.

This single-volume encyclopedia comprises 477 articles taken primarily from the fifth edition of *McGraw-Hill Encyclopedia of Science and Technology*, published in 1982. This would be a useful book for individuals as well as libraries that have a strong interest in electronics and would like to have an encyclopedia for patrons to borrow. The addition of new material is not enough reason to add this volume to a collection which already includes the *McGraw-Hill Encyclopedia of Science and Technology*.

106.   **McGraw-Hill Encyclopedia of Science and Technology.** 6th ed. New York: McGraw-Hill, 1987. 20v. index. LC 86-27422. ISBN 0-07-079272-2. $1,600.00/set.

This classic encyclopedia in science and technology has undergone massive revision since it was first published in 1960. It has been expanded from 15 to 20 volumes. Of the 7,700 articles, 2,000 are new or totally revised, many in the areas of electronics and communications. Areas specifically revised or added are computers, robotics, optical communications, and applications of lasers and electronic communication. One of the most noticeable changes is the use of a more modern design, including boldface type, two-color graphs, and a new typeface.

Each signed entry has a short bibliography. The encyclopedia is aimed at high school and college students, but is useful for graduate students who may need information outside of their field. All libraries of any size with any interest in science and technology should own this set.

# 10 Directories and Biographical Tools

In all fields there is often the need to quickly find biographical and directory-type information on individuals engaged in research or teaching. This is especially true in all areas of engineering because engineers are often involved in hiring consultants and buying products or equipment. As a result, engineers where ever they work need fast access to the names of people and companies involved in electrical and electronics engineering as researchers, teachers, manufacturers, or salesmen.

The following section lists directories containing biographical information along with those that list names and addresses of people and organizations, give company information, or give relevant marketing information. This section might just as well have been called "Guides to People and Places." While emphasis has been given to sources of current information, a few selected retrospective sources, which consist entirely of biographical information, are included since users frequently need biographical information on famous inventors, scientists, and researchers. Not included are lists of noncurrent products, services, and salesmen because they generally are of interest to lawyers involved in product liability cases and writers of company histories rather than engineers or students.

# Biographical Dictionaries and Directories

107. **American Men and Women of Science: Physical and Biological Sciences.** 16th ed. Edited by Jaques Cattell Press. New York: R. R. Bowker, 1986. 8v. illus. index. LC 76-7326. ISBN 0-8352-2221-7. $595.00.

Contains biographies of 127,000 people, 7,280 of whom are new to the 16th edition. Coverage is limited to well-known, living scientists in "the physical and biological fields as well as public health, engineering, mathematics, statistics and computer science." Also new in the 16th edition is a "discipline" index which organizes entrants by disciplines, which were adapted from the National Science Foundation's Taxonomy of Degree and Employment Specialities.

A cumulative index for editions 1-14, 1906-1979, is now available from R. R. Bowker (847p. LC 76-7326. ISBN 0-8352-1238-6).

108. **Biographical Dictionary of Scientists.** 3rd ed. Edited by Trevor I. Williams. New York: Halsted Press, 1982. 674p. illus. LC 82-160058. ISBN 0-470-27326-7. $29.95.

Originally this book consisted of short biographical entries of well-known, living scientist arranged in a dictionary format. Several major changes have been made in the third edition including the addition of 40 new entries, the addition of new footnotes to the original entries, the expansion of the chronological table, and, most importantly, the addition of a subject index. Consequently, it is now possible to access scientists by their area of specialty. Unfortunately the text was not reset and the new names were added as an addendum.

109. **Biographical Encyclopedia of Scientists.** Edited by Urdang Associates. New York: Facts on File, 1981. 936p. index. LC 80-23529. ISBN 0-87196-396-5. $80.00.

The 2,000 entries in this book include abbreviated biographies and discussions of scientific achievement. Emphasis has been placed on individuals who have contributed to the "pure sciences" of physics, chemistry, biology, astronomy, and earth sciences. Selected people who have made contributions to engineering and technology are also included.

110. **Dictionary of Scientific Biography.** Edited by Charles Coulston Gillispie. New York: Scribners, 1970-1980. 16v. including supplement. index. LC 69-18090. ISBN 0-684-14779-3. $55.00.

The first 14 volumes include 5,000 names selected by the editorial board and its consultants under the auspices of the American Council of Learned Societies. Volume 15 is a supplement covering additional biographies, while volume 16 contains the index. The controlling criterion for admission was whether the scientists' contributions "were sufficiently distinctive." Most entries include a short bibliography of works about and by the subject. While coverage is international in scope, most of the subjects are either European or American. Like most biographical tools, this one is biased toward prize-winners and those well known outside of their field.

111. **Directory of Consultants in Electronics.** Woodbridge, Conn.: Research Publications, 1985. 334p. index. ISBN 0-89235-097-0; ISSN 0882-6064. $85.00.

This is a directory of 2,726 self-designated consultants from the United States and Canada. The individuals were originally selected from *Who's Who in Technology Today*.

112.  **International Who's Who in Engineering.** 1st ed. Edited by E. Kay. London: International Biographical Centre, 1984. 550p. ISBN 0-900332-71-9. $115.00.

Provides biographical information on well-known engineers, with a heavy emphasis placed on European engineers. The intent is to update this on a regular basis. Future editions will only be available on standing order.

113.  **Leading Consultants in Technology.** 2nd ed. Woodbridge, Conn.: Research Publications, 1985. 2v. index. ISSN 0749-9000. ISBN 0-89235-089-x. $195.00/set.

A collection of biographical entries taken from *Who's Who in Technology Today*, with no additional material. Volume 1 includes biographical data organized by fields and by geographic location. The entries consist of traditional biographic data followed by a listing of important papers, patents, and honors. The fields are very broad. For example, entries under "Electronics" are further subdivided into electrical engineering, computer science, control systems, magnetics, and related electronic technologies. Volume 2 consists of "Index of Expertise" and "Index of Names." The expertise index is made up of words used by the biographee to describe his or her work. The name index is especially important because of the manner in which the entries are organized.

114.  **McGraw-Hill's Leaders in Electronics.** New York: McGraw-Hill, 1979. 651p. ISBN 0-07-019149-2. $47.50.

This somewhat dated work includes some 5,200 notables from private industry, government, academia, and the consulting and military worlds. The usual biographical information is given. An index is included which provides access to organizational affiliation.

115.  **Who's Who in Engineering.** 6th ed. New York: American Association of Engineering Societies, 1985. ISBN 0-87615-014-8. $200.00.

This directory, which is published every three years, consists of an alphabetical listing of living engineers with a baccalaureate or higher degree in engineering, who are registered in one or more states and are members of one of the engineering societies. A list of member societies is found in the front. An index of individuals by state and by specialization is provided.

116.  **Who's Who in Frontier Science and Technology.** 1st ed. Chicago: Marquis, Who's Who, 1984-1985. 842p. index. LC 82-82015. ISBN 0-8379-5701x. $84.50.

Consists of biographies of people who according to the publisher are "engaged in the forefront fields of science." The biographies, which are arranged in alphabetical order, include vital statistics, education, career history, awards, publications, and memberships along with a description of current research or work interests. An index is provided for the fields and subfields of technology. Engineering is broken down into large subfields including electrical, applied; electrical, computer engineering; and electrical, semiconductors.

117.  **Who's Who in Technology.** 2nd ed. Edited by Karl Strute. New York: UNIPUB, 1984. 3v. index. ISBN 3-921220-37-8. $120.00.

Comprehensive information on people and organizations in Western Europe. Covers more than 20,000 individuals and 650 companies in 26 technical fields.

118. **Who's Who in Technology Today.** 4th ed. Woodbridge, Conn.: Research Publications, 1984. 5v. index. ISSN 0190-4841. $425.00.

Provides access through 39 discipline categories, key word index, and an alphabetical list of 36,000 research scientists, engineers, and executives of high-technology companies. Individual volumes, each of which covers a specific subject area, may be purchased separately. Each entry, which consists of the traditional biographical information along with a description of the individual's work and a list of important papers, books, and honors, indicates whether the individual is willing to be a consultant or expert witness. Because entries are grouped by subfield, the alphabetical index is especially important.

## Membership Lists and Commercial Directories

119. **Computer Directory and Buyers' Guide.** Vol. 23- . Newtonville, Mass.: Berkeley Enterprises, 1974- . annual. ISSN 0010-4795. $27.00/yr. when purchased separately. (The October issue of *Computers and People*).

Consists of rosters and lists of companies and individuals in all areas of the computer industry. This is an excellent source for up-to-date information on firms, products and services, facilities, and hardware leasing.

120. **Electric World Directory of Utilities.** New York: McGraw-Hill, 1967- . annual. LC 16-21431. $225.00/yr.

Lists in alphabetical order the electric utilities in the United States, its possessions, and Canada. Each entry gives the officers, amount of power produced, sales address, and an abbreviated rate structure. The appendexes contain lists of Electric Reliability Councils and power pools, consulting engineers, and national and state associations.

121. **Electronic Industries Association Trade Directory and Membership List.** Washington, D.C.: Electronic Industries Association, 1933- . annual. ISSN 0091-9519. $50.00/yr. nonmember.

Consists of lists of industrial members who are involved in the manufacture of electronic products including consumer, industrial, and military electronics. A good source for addresses for some of the lesser known companies.

122. **Electronic Industry Telephone Directory.** Twinsburg, Ohio: Harris Publishing, 1963- . annual. ISSN 0422-9053. $35.00/yr.

A worldwide directory of companies involved in electronic sales and manufacturing. This, and its companion publication *Who's Who in Electronics*, are probably the best known and most used directories by sales people, electronics executives, procurement officers, librarians, students, and engineers.

123. **Electronic Market Data Book**. Washington, D.C.: Electronic Industries Association, 1951- . annual. ISSN 0070-9867. $55.00/yr.

Tabular listing of U.S. electronic product activity for the year. Includes production and sales statistics for U.S. markets.

124. **Electronic Marketing Directory.** New York: National Credit Office, 1959- . annual. $45.00/yr.

A marketing directory of all electronic manufacturers. Each listing includes plant locations, products, purchasing agent, dollar sales, and number of employees.

125. **Electronic News Financial Fact Book and Directory.** New York: Fairchild Publications, 1961- . annual. ISSN 0070-9875. ISBN 0-87005-469-4. $100.00/yr.

Financial profiles of over 700 publicly owned electronics corporations. Listings include sales, company size, number of employees, etc.

126. **IEE List of Members.** Piscataway, N.J.: IEE, 1984-1985. 592p. annual. $50.00/yr. nonmember.

Membership list for the Institution of Electrical Engineers, the English society for electrical engineers.

127. **IEEE Membership Directory.** Piscataway, N.J.: IEEE Press. annual. ISSN 0073-9146. $75.00/yr. nonmember.

Primarily a listing of the members of the largest American association of electrical and electronic engineers. Each entry provides address, job title, type of degrees and date, and work location. Of special interest are the short biographies of IEEE fellows, the lists of officers and directors of the IEEE, and a list of honorary members.

128. **National Society of Professional Engineers and Private Practice.** Washington, D.C.: NSPE, 1965- . annual. LC 68-29475. $45.00/yr.

A state-by-state, country-by-country directory of engineers in private practice who have passed the United States Professional Examination. Access is also possible by broad category, e.g. electrical. For those who do not belong to the NSPE, the same directory is now available from the society under the name *Professional Engineering Directory*.

129. **Who's Who in Electronics: Electronic Industry Data in Depth.** Twinsburg, Ohio: Harris Publishing, 1976- . annual. illus. ISSN 0734-8576. $75.00/yr.

Lists approximately 14,000 electronic firms, distributors, and representatives in both alphabetical and geographical sequences. Each entry gives company name, address, names of executives or key personnel, number of employees, products, major SIC code, annual sales volume, and the year the company was established. Known by many engineers as "The Harris Directory," it is a companion publication to the *Electronic Industry Telephone Directory*.

# 11  Dictionaries

Dictionaries are among the most commonly used reference books in all fields and engineering is no exception. The fast pace of electronic development, manufacturing, and application of electronics to other fields has encouraged the development of specialized dictionaries that reflect the changing definitions and new terms which regularly enter the vocabulary and literature of electrical, electronics, and computer engineering.

The language of the electrical and electronics engineering literature is primarily English, however foreign languages are represented to some degree in the literature. Universities and colleges with a large number of international students may find it helpful to have dictionaries that cover the first language of these students.

The following is a selected list of both definitional and polyglot dictionaries. They cover not only electrical and electronics engineering but the related fields of communications and computer science.

130.  **Communications Standard Dictionary.** By Martin H. Weik. New York: Van Nostrand Reinhold, 1982. 928p. illus. LC 81-21842. ISBN 0-442-21933-4. $42.50.

Defines terms in all areas of communications systems. As in Weik's other dictionary, *Fiber Optics and Lightwave Communications*, definitions are consistent with national, international, military, industrial, and scientific usage. The cross-references to synonyms and preferred terms are extensive. *Library Journal* called this dictionary "comprehensive."

131. **Computer Dictionary.** 3rd ed. By Charles J. Sippl and Roger J. Sippl. Indianapolis, Ind.: Howard W. Sams, 1984. 624p. illus. LC 79-91696. ISBN 0-672-21652-3. $15.95.

A straightforward dictionary which has gone through a number of printings. It is aimed at the new audience for computers, namely those managers and business people who have personal computers or who must communicate with the owners/operators of this equipment. The definitions cover all aspects of computers from programming to applications. An example of the variety covered can be demonstrated by the inclusion "multiaddress" and "document delivery." Each definition is fairly lengthy.

132. **Computer Glossary for Everyone: It's Not Just a Glossary.** 3rd ed. By Alan Freedman. Englewood Cliffs, N.J.: Prentice-Hall, 1983. 298p. illus. LC 82-62626. ISBN 0-13-164483-1. $14.95.

Entries consist of words frequently found in the popular personal computing literature and the definitions are aimed at nontechnical users. As a result proprietary names and trademarks are included as well as traditional words associated with computers, computing, data processing, and telecommunications. A list of software associated with the IBM PC is included.

133. **Dictionary of Computers, Data Processing, and Telecommunications.** By Jerry Martin Rosenberg. New York: John Wiley & Sons, 1984. 614p. illus. LC 83-12359. ISBN 0-471-885282-7. $15.85.

This book has definitions of over 10,000 words that relate either directly or indirectly to the usage of hardware and software. Each definition is followed by symbol, where relevant, representing the source for each definition. The appendix makes this publication particularly valuable because it lists equivalent terms in Spanish and French for those English terms defined in the body of the dictionary. Commonly used symbols, abbreviations and foreign phrases, and the term of choice in English are provided. The dictionary does an excellent job of giving the definitions of both very recent and obsolete terms. The definitions show the relationship that exists among the areas of computers, data processing, and telecommunications.

134. **Dictionary of Computing.** 2nd ed. Edited by Valerie Illingworth. New York: Oxford University Press, 1985. 430p. ISBN 0-19-853913-4. $29.95.

Containing over 3,750 terms, this dictionary was developed by 50 people working in the computer field. All aspects are covered, from formal languages and legal implications to hardware and memory devices. The entries range in complexity from basic ideas and equipment to graduate-level computer science terms and concepts. Terms are arranged in alphabetical order, with synonyms and common abbreviations following in brackets. Diagrams are provided as needed and adhere to standard conventions. This second edition was heavily updated.

135. **Dictionary of Electrical and Electronic Engineering: German-English, English-German.** 2nd ed. Edited by Hans F. Schwenkhagen. New York: International Publications, 1967. 909p. LC 67-74452. ISBN 3-7736-5072-8. $105.00.

A basic translating dictionary with an emphasis on German electrical and electronics terminology. Although it does not include some of the more recent electronic and computer engineering terms, this dictionary is still very useful for the basic vocabulary.

136. **Dictionary of Electrical Circuits: English and Chinese.** New York: French and European, 1975. 203p. ISBN 0-686-92288-3. $3.95.

A limited, somewhat dated dictionary covering words related to circuit design, development, and fabrication. It is inexpensive and one of the few English/Chinese electrical dictionaries currently available.

137. **Dictionary of Electrical Engineering.** 2nd ed. Edited by K. G. Jackson and R. Reinberg. Woburn, Mass.: Butterworth Publishers, 1981. 356p. LC 79-41721. ISBN 0-683-28354-2. $39.95.

Gives the definitions of words used in the United States and Great Britain by electrical engineers. A few definitions have illustrations and there are some cross-references. The definitions take into consideration usage differences between the United States and Great Britain.

138. **Dictionary of Electrical Engineering and Electronics: English-German.** Edited by P. K. Budig. New York: Elsevier Science Publishing, 1985. 720p. ISBN 0-444-99595-1. $65.00.

Prepared by a team of editors, the vocabulary in this dictionary is based on recent usage in specialized books, journals, and manuals. Over 60,000 terms are included, many of which have definitions. A German-English version is in preparation. This very up-to-date dictionary should be useful for a number of years to come.

139. **Dictionary of Electrical Engineering, Telecommunications and Electronics: English, German, French.** Wiesbaden, West Germany: O. Brandstetter WG, 1967. 3v. ISBN 0-9913001-4. $119.00. (Distributed by Heyden, Philadelphia).

A polyglot dictionary, with volume 1 covering German-English-French, volume 2 French-English-German, and volume 3 English-German-French.

140. **Dictionary of Electronics.** By Stanley William Amos. London: Butterworth Publishers, 1981. 335p. illus. LC 80-40233. ISBN 0-408-800331-6. $39.95.

See annotation for *Dictionary of Telecommunications* by S. J. Aries (151).

141. **Dictionary of Electronics.** Edited by E. C. Young. New York: Penguin, 1979. ISBN 0-14-051074-8. $5.95.

A paperback dictionary aimed at beginning undergraduates.

142. **Dictionary of Electronics: English-German.** By Alfred Oppermann. Detroit, Mich.: Gale Research, 1980. 692p. ISBN 3-598-10312-3. $75.00.

A basic translation dictionary.

143. **Dictionary of Electronics, Communications and Electrical Engineering.** Edited by H. Wernicke. New York: Adlers Foreign Books, 1980. 2v. ISBN 0-685-05199-4, 0-685-05200-1. $72.00/set.

A 1,300-page dictionary covering all aspects related to electrical and electronics engineering. Some emphasis on European usage.

144. **Dictionary of Engineering and Technology.** 4th ed. Edited by Richard Ernst. New York: Oxford University Press, 1980. ISBN 0-19-520269-4. $69.00.

A scholarly dictionary with a strong emphasis on European usage.

145. **Dictionary of Logical Terms and Symbols.** By Carol Horn Greenstein. New York: Van Nostrand Reinhold, 1978. 188p. illus. ISBN 0-442-22836-8. $10.95.

The various notational systems used by logicians, engineers, and computer scientists are covered in this dictionary. The multicolumn tables enable translations from one notational system to another and present the many ways in which key English expressions can be formalized. The main symbolic forms of Boolean algebra; set theory; quantification theory; two-termed relations; logic gates; and epistemic, doxastic, deontic, and tense logical are presented in alternative systems, such as Peano-Russel, Hilbert, and Polish notations, as well as in English. It also includes program flowchart symbols, consistency trees, truth-functional and binary truth tables, and 66 circuit diagrams.

146. **Dictionary of Microelectronics: English/German and German/English.** By W. Bindmann. New York: Elsevier Science Publishing, 1984. 489p. ISBN 0-444-99619-2. $109.75.

This bilingual dictionary, which covers 22,000 terms in German and English, was designed to be used by researchers and writers in both languages. Coverage includes microelectronic hardware, software, and devices; solid-state physics; microlithography; and peripheral equipment. The inclusion of fabrication and application terms would increase the usefulness to translators.

147. **Dictionary of Microprocessor Systems.** Edited by D. Mueller. New York: Elsevier Science Publishing, 1983. 320p. ISBN 0-444-99645-1. $74.50.

A multilingual dictionary containing 10,000 terms in English, German, French, and Russian from the literature of microprocessor technology, computer technology, and programming.

148. **Dictionary of Printed Circuit Technology: English-German, German-English.** Edited by Guenther Boehss. New York: Elsevier Science Publishing, 1985. 185p. LC 85-12955. ISBN 0-444-99555-2. $53.75.

This dictionary lists about 6,000 terms in English and in German. It covers all aspects of printed circuit technology, from instrument wiring to the materials of construction.

149. **Dictionary of Scientific and Technical Terminology, Five Lingual: English, German, French, Dutch, Russian.** By. A. S. Markow and V. Romanov. Hingham, Mass.: Kluwer Technical Books, 1984. 496p. ISBN 0-318-01661-3. $34.50.

This polyglot dictionary contains approximately 9,000 entries and covers all aspects of science and technology that might be needed by an engineer, including basic terms in electrical engineering. The dictionary was compiled by Kluwer Technical Books in conjunction with Russian Language Publishers.

150. **Dictionary of Technology: English-German and German-English.** Edited by Rudolf Walther. New York: Elsevier Science Publishing, 1985. 2v. ISBN 0-444-99491-9, 0-444-99590-0. $129.95/ea.

A comprehensive dictionary covering all areas of science, technology, and industrial production. Over 100,000 terms are defined. An appendix contains important abbreviations.

151. **Dictionary of Telecommunications.** By S. J. Aries. London: Butterworth Publishers, 1981. 336p. illus. LC 80-41449. ISBN 0-408-00328-6. $39.95.

Associated with the *Dictionary of Electronics* by S. W. Amos. The subjects of both books are closely related and some common concepts appear. The books are designed to be self contained but detailed explanations of terms are confined to the *Dictionary of Electronics.* Both books attempt to cover American and European usage. Terms defined elsewhere in the text appear in bold italics. The appendixes include common abbreviations and acronyms. Those marked with an asterisk are defined in the dictionary.

152. **Electronics Dictionary.** 4th ed. By John Markus. New York: McGraw-Hill, 1978. 745p. illus. LC 77-13876. ISBN 0-07-040431-3. $39.50.

By the well-known and prolific electronics writer, this dictionary defines 17,000 terms. It is heavily illustrated with diagrams and drawings chosen to clarify the more complex definitions. Covers all areas of electronics including the related areas of lasers, microcomputers, video games, and satellite communications. One difference between this and other dictionaries is the inclusion of an electronics style manual. The author has elaborated on how electronics drawings, schematics, and exaplanations should be handled.

153. **Electro-Optical Communications Dictionary.** Edited by Dennis Bodson and Dan Botez. Rochelle Park, N.J.: Hayden Book Co., 1983. 168p. illus. LC 83-4295. ISBN 0-8104-0961-5.

This comprehensive listing of terms and definitions associated with fiber optics and lightwave communications systems includes 25,000 entries that are consistent with international, federal, industrial, and technical society standards. The dictionary is aimed at all levels ranging from manufacturers to the users of fiber optic instruments. A list of acronyms and abbreviations is included.

154. **Elsevier's Dictionary of Opto-Electronics and Electro-Optics.** Edited by K. Nentwig. New York: Elsevier Science Publishing, 1986. 295p. LC 86-4427. ISBN 0-44442617-5. $101.75.

From the well-known publisher of scientific and technical polyglot dictionaries, this text covers prosaic words and phrases like energy loss, garnet, or polarized, and more importantly new vocabulary like epitaxial layer, excimer laser, and terms for the III-V groups of semiconductor materials.

155. **Encyclopedic Dictionary of Electronic Terms.** By Robert E. Traister. Englewood Cliffs, N.J.: Prentice-Hall, 1984. 608p. illus. LC 84-2107. ISBN 0-13-276990-0. $29.95.

"A quick source for the serious comprehension of the basics involved in electronics. Each topic has been chosen and its explanation written to aid the beginner as well as the experienced technician" (preface). Organized in alphabetical order, each entry is at least twice as long as those in most dictionaries. There is heavy use of illustrations with many material and electric properties given in tabular format.

156. **English-Chinese Glossary of Electronic and Electrical Engineering.** New York: French and European, 1980. 636p. ISBN 0-686-97364-x. $29.95.

One of the few up-to-date English-Chinese dictionaries currently available. It is especially strong in electrical and power terminology.

157. **English-German, German-English Solid State Physics and Electronics Dictionary.** By G. Birdmann. London: Collets', 1980. 1,103p. ISBN 0-569-07204-2. $100.00. (Distributed by State Mutual Book).

158. **English-Spanish, Spanish-English Encyclopedic Dictionary of Technical Terms.** Edited by Javier L. Collazo. New York: McGraw-Hill, 1980. 3v. ISBN 0-07-70172 (English); 0-07-079162-7 (Spanish). $154.00.

This is a basic translation dictionary.

159. **Facts on File Dictionary of Telecommunications.** By John Graham. New York: Facts on File, 1984. 224p. illus. LC 81-15675. ISBN 0-87196-876-2. $15.95.

Aimed at both the North American and European audience, the terms defined in this dictionary are in "keeping with international practice." A nice touch is that words used within definitions that are defined elsewhere in the dictionary are italicized. Appendixes include longer discussions, diagrams and examples of some basic concepts such as pulse amplitude modulation.

160. **Fiber Optics and Lightwave Communications: A Standard Dictionary.** By Martin H. Weik. New York: Van Nostrand Reinhold, 1980. 320p. illus. LC 80-12765. ISBN 0-442-25658-2. $22.95.

The author states that "the terms were taken from the spoken words of designers, developers, researchers, operators, educators and from the literature." Definitions are given for many traditional words like "bandwidth" and "waveguide" and many computer terms like "gate" and "half duplex circuits." Words defined elsewhere in the text are in italics. An extensive bibliography is included.

161. **IEC Multilingual Dictionary of Electricity.** By IEEE. New York: Wiley-Interscience Publication, 1984. 461p. ISBN 0-471-80784-2. $49.95.

English words and the definition are listed first, with the corresponding terms in French, Russian, German, Spanish, Italian, Dutch, Polish and Swedish following. Defines over 7,500 terms in English. A list of common expressions under each term clarifies variant meanings. Each definition was approved by scientists and engineers from each of the countries included.

162. **IEEE Standard Dictionary of Electrical and Electronics Terms.** 3rd ed. Edited by Frank Jay. New York: Wiley-Interscience Publication, 1984. 1,173p. LC 84-81283. ISBN 0-471-890787-7. $55.00. (IEEE/ANSI STD 100-84).

Unlike other dictionaries, the *IEEE Standard* includes the source for each definition, most of which were IEE, IEEE, ANSI, or ASTM standards. Terms are heavily cross-referenced. Multiple meanings for different areas of electronics are given. The appendix contains an extensive list of abbreviations, symbols, acronyms, functional designations, and letter combinations. One problem associated with this dictionary is that a term must have appeared in a standard in order to be included. This has been enlarged and heavily updated since the second edition was published in 1977.

163. **Illustrated Dictionary of Electronics.** 3rd ed. By Rufus P. Turner and Stan Gibilisco. Blue Ridge Summit, Pa.: Tab Books, 1985. 595p. illus. LC 85-4623. ISBN 0-8306-0866-4. $34.95.

Completely revised and updated since the 1982 third edition, this basic dictionary is aimed at all levels of users in all areas of electronics. Over 27,000 words, acronyms, and abbreviations are defined, many illustrated with clear drawings. Unlike some other dictionaries, this one gives excellent coverage of abbreviations. Although the definitions are quite short, they are clear and helpful. In addition to electronics, coverage includes all aspects of the traditional areas of electrical engineering as well as the related areas of communications and computers. Appendixes include symbols, conversion tables, and resistor codes.

**164. Illustrated Dictionary of Microcomputer Terminology.** By Michael Hordeski. Blue Ridge Summit, Pa.: Tab Books, 1978. 322p. illus. LC 78-10421. ISBN 0-8306-1088-x. $10.95.

A dictionary of 4,000 computer terms, business words, and jargon with short definitions. Attempts to reflect current usage of hardware and software terms, with particular emphasis on software.

**165. Illustrated Dictionary of Microelectronics and Microcomputers.** By R. C. Holland. New York: Pergamon Press, 1985. 168p. illus. ISBN 0-08-0316352. $30.00.

Terms are presented alphabetically with cross-references and illustrations where necessary. The stated aim is to define and explain new terminology for expressions in common use by workers and to describe new electronic devices, systems, and programming techniques. Definitions are given for a number of manufacturing terms like "mask" and "photo tape up," which are found in trade literature but uncommon in dictionaries. All recent circuits, systems, and applications are included.

**166. Illustrated Electronics Dictionary.** Edited by Howard Berlin. New York: Merrill, 1986. 188p. LC 85-62113. ISBN 0-675-20451-8. $12.95.

A basic electronics dictionary aimed at the hobbyist and technician. Contains limited physics, math, or theoretical vocabulary.

**167. INSPEC Thesaurus.** Hitchin, Herts., England: Institution of Electrical Engineers, 1973- . annual. ISBN 0-85296-255-8. $25.00/yr.

A basic hierarchical thesaurus with over 10,000 terms, 5,500 preferred terms, and 4,500 cross-references. There are no definitions. For use with the INSPEC database and indexes.

**168. McGraw-Hill Dictionary of Electrical and Electronic Engineering.** Edited by Sybil P. Parker. New York: McGraw-Hill, 1984. 488p. LC 84-28837. ISBN 0-07-045413-2. $15.95pa.

A small dictionary which is duplicated almost entirely by the other McGraw-Hill dictionaries. Little or no coverage is given to the related areas of electrical engineering like communications nor are abbreviations well covered. Abbreviations are rampant in electronics so this is a real limitation. Also there are no illustrations. There are other better dictionaries on the market.

**169. McGraw-Hill Dictionary of Electronics and Computer Science.** New York: McGraw-Hill, 1984. 592p. LC 83-20362. ISBN 0-07-045416-7.

Electronics and computer science definitions were taken directly from the *McGraw-Hill Dictionary of Scientific and Technical Terms* so as to be sold as specialty dictionary. There is no need to buy or own this dictionary if you already have the more comprehensive *Dictionary of Science and Technology.*

170. **McGraw-Hill Dictionary of Science and Engineering.** Edited by Sybil P. Parker. New York: McGraw-Hill, 1984. 960p. illus. LC 83-18265. ISBN 0-07-0454833. $32.50.

The 36,000 terms defined in this dictionary were selected from the 1984 edition of *McGraw-Hill Dictionary of Scientific and Technical Terms.* The definitions include some abbreviations, symbols, and both SI and customary units.

171. **Modern Dictionary of Electronics.** 6th ed. By Rudolf F. Graf. Indianapolis, Ind.: Howard W. Sams and Co., 1984. unpaged. illus. LC 83-51223. ISBN 0-672-22041-5. $39.95.

Defines 20,000 terms that are unique to electronics and related fields. Including communications, reliability, semiconductors, microelectronics, fiber optics, medical, and computer electronics.

172. **Russian-English Polytechnical Dictionary.** Edited by B. V. Kuznetsov. Moscow: Russian Language Publishers, 1980. 723p. ISBN 0-8285-1851-3. $30.00.

Covers over 90,000 Russian words with their English meanings and equivalents.

173. **Scientific Terms Electrical Engineering (English-Japanese).** By Ministry of Education, Science, and Culture. New York: French and European, 1984. 675p. ISBN 0-686-97433-6. $39.95.

A modern translation dictionary covering a language which is becoming very important in electronics.

174. **Semiconductors International Dictionary in Seven Languages.** Edited by D. Dunster. Boston: Cahners Publishing, 1971. 2v. illus. index. (International Dictionaries of Science and Technology).

An impressive compilation of terms from all phases of semiconductor usage. Each term or concept is briefly defined, with an added illustration or table as necessary. Entries are arranged alphabetically by the English term, with the corresponding German, French, Italian, Portuguese, Russian, and Spanish.

175. **Technik-Wörterbuch: Elektronik, Electrotechnik.** London: State Mutual Book, 1980. 450p. ISBN 0-6896-7209-1. $120.00.

A basic German English dictionary of electronics and electrotechnology.

# 12    Data Compilations

Books listed in this section are primarily listings of numeric data, presented in a concise, often tabular format. The objective is to provide quick access or lookup to the data themselves, rather than an explanation of the phenomena or description of the calculation or experiment used to obtain the data. The data could be physical constants; chemical, electrical or mechanical properties; characterizations of materials; and parameters and major characteristics of electrical and electronic devices and equipment. Types of data typically found are thermal expansion coefficients for various materials, power requirements of equipment, wave characteristics, etc.

Many of the books listed in the handbook section may also contain data. What differentiates books found in this section from those in other sections is the presentation. Books found here are concerned with data and generally contain only limited explanations, while books listed in the handbook section have extensive written explanations or commentaries. The data presented in the handbook section are often illustrative, where as data presented in the following books are presented for their own sake. This type of tabular presentation is very popular with engineers, librarians, and students, who are often looking for just one particular piece of datum. Because this type of presentation is so popular, the following section separates this type of material.

176.   **Butterworth and Chebyshev Digital Filters: Tables for Their Design.** Amsterdam: Elsevier Science Publishing, 1973. unpaged. no index. LC 73-89154. ISBN 0-444-41178-x.

Tabular presentation of both Butterworth and Chebyshev digital filter constants. Four worked examples of filter design show how to use the book. These tables may also be used to design highpass filters.

√ 177. **Contemporary Electronics Circuits Deskbook.** Compiled by Harry L. Helms. New York: McGraw-Hill, 1986. 253p. illus. index. LC 85-19721. ISBN 0-07-027980-2. $29.95.

A compilation of circuit designs and applications which have appeared in recent electronics magazines, application notes, and databooks. The contents includes active filters, digital circuits, LED and optoelectronic circuits, repeater circuits, and voltage regulation circuits. Each diagram was reproduced directly from the original and includes type or part numbers, values, identifying title, and citation of the original source.

178. **D.A.T.A. Books: Electronic Information Series.** San Diego, Calif.: D.A.T.A. Inc., 1978- . 45 times/yr. ISSN 0732-5894. $1,661.00/yr. (set).

This massive accumulation of information on electronic devices produced throughout the world consists of 28 books, the aim of which is to facilitate device selection or, when relevant, device substitution. Each book covers a specific type of device. Generally users of this book would know what type of device they were using, and therefore what book to go to. The usefulness of the set, however, has been enhanced by the addition of a two-volume reference set which covers all the volumes. These references may be purchased separately, or are complimentary with purchase of the complete set.

The books are available in three general areas: integrated circuits, discretes and special applications, and discontinued devices. The following annotations for the individual books are grouped by these areas. These books are not for all libraries; they are intended for very large comprehensive university collections or collections associated with manufacturers or designers.

*D.A.T.A. Reference Volumes*

**Application Notes Reference.** ISSN 0090-3655. $43.00/yr.
Covers the order and application notes for 4,200 circuits and devices from 96 manufacturers. Published semi-annually. Formerly *Semiconductor Application Notes.*

**Master Type Locator.** ISSN 0730-6776. $51.00/yr.
Master index for the *D.A.T.A. Books.* Identifies manufacturers and product classes for each type and then directs the user to the specific *D.A.T.A. Book* where detailed technical data is given. Published annually.

*D.A.T.A. Books Covering Integrated Circuits*

**Audio/Video ICS.** ISSN 0276-5101. $43.00/yr.
Provides detailed electrical, functional, and pictorial information on 2,600 television circuits, audio amplifiers, clock circuits, calculators, rhythm devices, and video games from 52 manufacturers. Published semi-annually.

**Digital ICS.** ISSN 0193-4295. $74.00/yr.
References to more than 14,000 ICs from 52 worldwide manufacturers arranged by primary device parameters. Includes basic logic, timing parity and latch functions, and thousands of logic and outline drawings. Also included are 1,565 military-qualified types. Published semi-annually. Formerly *Digital Logic/Computational Integrated Circuit D.A.T.A. Book.*

**IC Functional Equivalence Guide.** $74.00/yr.

References over 20,000 memory, digital, and interface devices from over 200 manufacturers. Equivalence is determined by an independent laboratory. Published semi-annually.

**Interface ICS.** ISSN 016-0119. $74.00/yr.

Electrical, logical, physical, and connection data on more than 11,900 state-of-the-art devices from 88 worldwide manufacturers. Arranged by device parameters within functional groupings. Logic/peripheral drivers, level converters, receivers, and specialty sensors are also included. There are logic and outline drawings for each type listed. Published semi-annually.

**Linear ICS.** ISSN 0270-9988. $74.00/yr.

Provides characteristics for over 8,800 IC types from 79 manufacturers in the following categories: operational, differential, audio, wideband and RF/IF amplifiers, and voltage regulators. Includes schematic, logic, and outline drawings for each type. Published semi-annually. Formerly *Linear Integrated Circuit* (ISSN 0024-3809).

**Memory ICS.** ISSN 0195-5853. $74.00/yr.

A reference to chip level products (6800) that do processing. Includes processor architecture and manufacturers' software support. Drawings, manufacturer's names, and addresses are cross-referenced. Published semi-annually. Formerly *MSI-LSI Memory* (ISSN 016309226).

**Microcomputer Systems.** ISSN 0276-5098. $69.00/yr.

Covers 3,700 devices which are organized by common denominator at design level, for example, bus structure, CPU, and controller. Special features include explanations of bus structure and the organization of support boards. Also includes combined physical and block drawings. Published semi-annually.

**Modules/Hybrids.** $74.00/yr.

Details over 9,900 linear and interface modules and hybrid devices from 71 worldwide manufacturers. Includes over 6,000 power devices. Published semi-annually.

**Power Supplies.** $74.00/yr.

Cross-references more than 23,900 linear and switching power supplies and DC/DC converters from 48 worldwide manufacturers. All types listed by volts in, amps in, volts out, amps out, operating temperature ranges, general descriptions and more. Published semi-annually.

*D.A.T.A. Books Covering Discretes and Special Applications*

**Diodes.** ISSN 0271-0803. $69.00/yr.

Contains information on 46,600 diodes from 157 manufacturers. Includes dimensioned outline drawings for most diode types. Published annually.

**Microprocessor Software.** ISSN 0276-511x. $69.99/yr.
Details 2,800 packages from 164 vendors with descriptive vendor details. Covers resident, cross systems, system support, software, and engineering application. Package information includes language, peripherals required, and memory requirements. Published semi-annually.

**Microwave.** ISSN 0271-0773. $69.00/yr.
Provides information on 22,100 semiconductors and tubes from 109 manufacturers. Includes source, amplifier, output and duplexer tubes, mixer, detector, varactor, tunnel, and PIN and oscillator diodes. Includes military specification numbers and dimensioned outline drawings. Published annually.

**Optoelectronics.** ISSN 0164-002x. $69.00/yr.
Covers 17,500 devices in 22 sections with the assemblies arranged by primary device parameters and grouped as follows: emitters, junction sensors, photocells, photocouplers, displays and specialty devices, and fiber optics components. Schematic and outline drawings are included. Published annually.

**Power Semiconductors.** ISSN 0164-0038. $74.00/yr.
Some 57,200 power devices from 157 worldwide sources are detailed. Includes standard and fast recovery rectifiers, power transistors, general purpose and inverter SCRs, plus package outline drawings and lead code identification. Published annually.

**Thyristors.** ISSN 0732-6092. $74.00/yr.
SCRs, Triacs, Schockley diodes, gate turn-off devices, SCS, and triggers from 79 manufacturers are covered. Also includes 41,500 types with lead designations and dimensioned outline drawings. Published annually.

**Transistors.** ISSN 0732-6203. $69.00/yr.
Lists over 30,000 type numbers from 133 worldwide manufacturers arranged in 14 technical data sections. Coverage includes a special military specification index and dimensioned outline drawings. Published annually.

*D.A.T.A. Books Covering Discontinued Devices*

**Discontinued Digital and Audio/Video.** $51.00/yr.
A complete index, with drawings, of over 19,200 devices that have become obsolete since 1968. Designed to be used as a cross-reference guide from obsolete to active devices. Published annually.

**Discontinued Diodes.** ISSN 0270-9465. $43.00/yr.
Lists 32,300 diodes obsolete since 1969. Published annually.

**Discontinued Interface and Memory.**
This referencing system indexes over 14,900 devices, with drawings, that have become obsolete since 1970. Sections include logic buffers/drivers, line drivers/transmitters, peripheral/power drivers, A/D and D/A converters, etc. Published annually.

**Discontinued Linear.** $51.00/yr.

Facilitates substitution when used with *Linear D.A.T.A. Book.* Lists 5,800 types obsolete since 1969. Published annually.

**Discontinued Microwave.**

Technical data on over 18,100 devices that have become obsolete since 1964. Designed to be used with *Microwave D.A.T.A. Book.* Published annually.

**Discontinued Optoelectronics.** ISSN 0732-4235. $51.00/yr.

Lists over 7,600 devices obsolete since 1974. Designed to be used with *Optoelectronics* when looking for a substitute. Published annually.

**Discontinued Thyristors.** ISSN 0092-508x. $43.00/yr.

Lists more than 24,700 devices, including SCRs and PNPNs discontinued since 1963 as well as specialty thyristors discontinued since 1973. Published annually.

**Discontinued Transistors.** ISSN 0730-4846. $43.00/yr.

More than 15,500 types obsolete since 1956 are listed. Published annually.

**Discontinued Type Locator.** ISSN 0730-4943. $51.00/yr.

Covers 140,700 obsolete ICs, transistors, diodes, thyristors, microwave, optoelectronic, and semiconductor devices. Designed to simplify substitution and replacement. Published annually.

179. **Handbook of Practical Electronic Circuits.** By John D. Lenk. Englewood Cliffs, N.J.: Prentice-Hall, 1982. 352p. illus. index. LC 81-227. ISBN 0-13-380741-x. $24.95.

Describes 270 circuits commonly used by hobbyists, designers, and students. The circuits cover all areas of electronics, audio frequency circuits, attenuators, operational amplifiers, and switching circuits. The book details circuit design, how the circuit performs, why it performs, and general characteristics.

180. **IC Master.** Garden City, N.J.: United Technical Publication, Herman Publishing, 1962- . annual. illus. index. LC 77-647487. $82.50/yr.

This is one of the better known compilations of integrated circuit data. The two-volume 1986 edition (ISBN 0-89047-0480) has over 12,000 device changes from the 1985. Volume 1 covers microprocessors, microcomputer boards, MPU development systems, digital, military, and applications notes. Coverage in volume 2 includes memory, interface, linear, and custom ICs.

The indexes include: part number and part guide, military parts index, advertisers index, and an application note directory. The latter provides a 20-30 word abstract of the note with references to the relevant manufacturers catalog or data sheet. Probably the most heavily used part of the book is the *Master Section Guide.* This section allows identification of ICs which match specific design requirements. Other listings include a directory with the addresses and telephone numbers of distributors, manufacturers, and representatives. Quarterly updates are available separately or as part of the subscription.

181.  **IC Schematic Sourcemaster.** By Kendall Webster Sessions. New York: John Wiley, 1978. 557p. illus. index. LC 77-13404. ISBN 0-471-02623-9. $56.95.

Contains 1,500 schematic diagrams for electronic circuits built around integrated packages. The book is organized into broad classes like voltage, current and power sources, television, and timers. Each circuit is identified by part number and company name. Each diagram was reproduced from original company data sheets. References are given to the original data.

182.  **Illustrated Encyclopedia of Solid-State Circuits and Applications.** By Donald R. Mackenroth and Leo G. Sands. Englewood Cliffs, N.J.: Prentice-Hall, 1984. 353p. illus. index. LC 83-23077. ISBN 0-13-450537-9. $29.95.

Heavily illustrated with schematics. Each semiconductor and basic circuit is explained and the characteristics are given in full practical detail. The stated aim of this book is to present a clear understanding of the principles and applications of solid-state circuitry so that the user can specify and use these devices.

183.  **Illustrated Encyclopedic Dictionary of Electronic Circuits.** By John Douglas-Young. Englewood Cliffs, N.J.: Prentice-Hall, 1983. 444p. illus. index. LC 82-23067. ISBN 0-13-450734-7. $29.95.

This book is divided into three parts: an illustrated encyclopedic dictionary of electronic circuits; designing, breadboarding, and final assembly; and a guide to performance testing and troubleshooting. An appendix includes schematic symbols, electronic formulas, electronic properties of commonly used materials, etc. The entry for each circuit includes a schematic and parts lists. The many illustrations use a step-by-step approach and have captions which are written clearly and concisely. This is a useful addition to most reference shelves.

184.  **Illustrated Guidebook to Electronic Devices and Circuits.** By Frederick W. Hughes. New York: Prentice-Hall, 1983. 432p. illus. index. LC 81-23544. ISBN 0-13-451328-2. $30.95.

Intended for the library of anyone interested in electronics, from the hobbyist to the engineer. The glossary consists of commonly used or encountered microprocessor terms along with basic theory, components, devices, circuits, and systems used in electronics. Chapter topics include optoelectronic devices, vacuum tubes, power supplies, amplifiers, oscillators, and solid-state devices.

185.  **Master Handbook of Microprocessor Chips.** By Charles K. Adams. Blue Ridge Summit, Pa.: Tab Books, 1981. 378p. illus. index. LC 80-28678. ISBN 0-8306-9633-4. $18.95.

A compilation of data on the popular types of microprocessors and some of their support chips. The operation of each chip is covered in detail with instruction sets. Organized by family, each description includes manufacturers, pin descriptions, and configurations. Most of the text is in outline format under large headings like function pin description or MPU addressing modes. Schematic diagrams are included.

186.  **Microprocessor and Microcomputer Data Digest.** By W. H. Buchsbaum and Gina Weissenberg. Reston, Va.: Reston Publishing, 1983. 336p. illus. index. LC 82-21620. ISBN 0-8359-4381-x. $28.95.

Covers four-, eight-, 12-, and 16-bit CPUs, as well as eight-, and 16-bit microcomputer chips. The book provides the essential technical data for each IC, functional block diagrams, and references to similar ICs. Each entry contains pin connection diagrams, descriptions for each pin connection, functional descriptions, and block diagrams for each circuit.

187. **Modern Electronic Circuits Reference Manual.** By John Markus. New York: McGraw-Hill, 1980. 1,238p. illus. index. LC 79-22096. ISBN 0-07-040446-1. $74.95.

Designed for ready reference, this compilation of data gathered from electronic application notes and journals describes over 2,630 modern electronic circuits. Each circuit description is complete with values for all parts. Performance details are arranged in 103 logical chapters. Citations are included at the end entry. Examples of entries include audio control circuits, burglar alarms, clock signal and fiber optic circuits, and memory and medical circuits.

188. **Properties of Amorphous Silicon: Handbook of Evaluated Data.** From EMIS Database. London: IEE, 1985. 260p. index. $195.00. (EMIS Datareviews Series, no. 1).

A collection of evaluated data on the properties of amorphous silicon, this handbook provides data for material prepared by the many deposition techniques including the influence of hydrogren, dopants, and deposition parameters. The text consists of 260 pages of numeric values, tables, and 130 data reviews (evaluative surveys). The material for this book has come from the EMIS database and is designed to aid in the design of more efficient and reliable large-area devices such as thin-film integrated circuits and image sensors.

189. **Properties of Gallium Arsenide.** London: IEE, 1985. $125.00. (EMIS Datareviews Series, no. 2).

A compilation of the best data available on the properties of gallium arsenide, GaAs. Data which are from the EMIS database and experts who have cooperated with this database, have been evaluated and reviewed by these experts. This compilation is for libraries and researchers who are actively engaged in semiconductor material research.

190. **Schaum's Outline of Theory and Problems of Basic Circuit Analysis.** By John O'Malley. New York: McGraw-Hill, 1982. 400p. illus. index. LC 80-26925. ISBN 0-07-047820-1. $7.95. (Schaum's Outline Series).

Like all the other outlines in this series, it covers in clear concise descriptions basic circuit analysis.

191. **Semiconductor General Purpose Replacements.** 5th ed. By Howard W. Sams engineering staff. Indianapolis, Ind.: Howard W. Sams, 1984. illus. index. ISBN 0-672-22418-6. $19.95.

A guide to general-purpose replacements for almost 225,000 bipolar and field-rectifiers, ICs, and more. Listed by U.S. and foreign type number and manufacturer's part number. Much of the data presented here come from data developed for the PHOTOFACT service.

192.  **750 Practical Electronic Circuits.** Edited by Roland S. Phelps. Blue Ridge Summit, Pa.: Tab Books, 1983. 576p. illus. index. LC 82-5988. ISBN 0-8306-2499-6. $21.95.

With the availability of sophisticated and highly technical components at the local electronics store devices like test equipment, multivibrators, detectors, and wave form and function generators, audio and visual circuits unknown only a few years ago now can be easily made by an amateur. Representative circuits are included from simple ones to those using modern integrated circuits. The diagrams are grouped by generic types and indexed by application circuits.

193.  **Special Circuits Ready Reference.** By John Markus. New York: McGraw-Hill, 1982. 230p. illus. index. LC 82-105. ISBN 0-07-040461-5. $12.50.

Portions of the book appeared in *Modern Electronic Circuits Reference Manual*, which is one of five electronics reference books by the same author. The information given is a synthesis of an extensive literature search of U.S. and foreign electronic journals and manufacturers information. Citations to the original work are given. Each circuit description includes type numbers, values of all significant components, performance data, and applications.

194.  **Tables for Active Filter Design.** By Mario Biey and Amedeo Premoli. Dedham, Mass.: Artech House, 1985. various paging. LC 84-073279. ISBN 0-89006-159-9.

Provides clear, well-documented filter tables for electrical engineers and researchers interested in data pertaining to CAUER and MCPER functions. Part I contains essential theory and frequency transformation charts. Part II, which entails over two-thirds of the volume, includes data on poles and zeros of CAUER and MCPER low-pass functions. This book would be a useful addition to high-level electrical engineering collections at universities and research facilities.

195.  **Tables of Antenna Characteristics.** By R. W. P. King. New York: Plenum, 1971. 400p. no index. LC 74-157425. ISBN 0-306-65154-8. $75.00.

Tabular presentation of the properties or characteristics of cylindrical dipoles, monopoles, circular loops, broadside and endfire arrays, and two element arrays antennas.

196.  **Towers' International Digital IC Selector.** By Thomas D. Towers. Blue Ridge Summit, Pa.: Tab Books, 1983. 260p. ISBN 0-8306-0616-5. $19.95.

An essential reference tool for anyone involved in the design and repair of digital equipment. Consists of quick look-up tables, which allow users to chose equivalent ICs easily.

197.  **Towers' International Microprocessor Selector.** By T. D. Towers. Blue Ridge Summit, Pa.: Tab Books, 1982. 160p. LC 81-18347. ISBN 0-8306-1716-7. $19.95.

Part of the authors series of selector books, this volume covers microprocessors and related devices from the United States, Europe, United Kingdom, and Japan in a tabular format. Data are given for 7,000 commercially available microprocessor chips and LSI circuits including device number, function application, description, family reference, manufacturer, and DC supply voltage. The appendixes provide a bibliography of references since 1975 as well as a glossary of terms.

198. **Towers' International Op Amp Linear-IC Selector.** By Thomas D. Towers. Blue Ridge Summit, Pa.: Tab Books, 1979. 190p. ISBN 0-8306-9771-3. $12.95.

Data bank of quick look-up tables, with easy-to-use glossaries and a useful introductory section. The major part of the book is organized by type number with the ratings, characteristics, case details, terminal identification, applications use, manufacturer, and substitution equivalents presented in tabular format.

199. **Towers' International Transistor Selector.** 3rd ed. By T. D. Towers. Blue Ridge Summit, Pa.: Tab Books, 1982. 280p. illus. LC 81-182484. ISBN 0-8306-1416-8. $19.95.

A tabular presentation of the basic specifications for 20,000 transistors from the United States, Europe, United Kingdom, Japan, and Russia. Information provided includes ratings, characteristics, case details, terminal identifications, applications, manufacturers, and substitution equivalents. Tables are supported by appendix materials on lead/terminal identification diagrams, case outline diagrams, house codes, CV transistor and related prototypes, and manufacturers listings.

200. **Transistor Specification Manual.** 7th ed. Indianapolis, Ind.: Howard W. Sams, 1975. 220p. illus. LC 75-5418. ISBN 0-672-21208-0. $21.95.

Arranged in three sections. Section 1 is arranged by specification number and gives voltage, power, current, temperature limits, polarity, leakage, gain, frequency parameters, and manufacturers. Section 2 is a guide to lead and terminal identification. Section 3 shows outlines of registered and nonregistered transistors.

201. **Transistor Substitution Handbook.** 17th ed. Indianapolis, Ind.: Howard W. Sams, 1978. illus. index. ISBN 0-672-21515-2.

A data compilation of American and foreign transistors arranged in both numerical and alphabetical order. Also listed are keys to manufacturers abbreviations, descriptions of types, polarity, and applications information. This volume was compiled to help the designer choose replacement transistors for existing devices.

202. **Tube Substitution Handbook.** 21st ed. By Howard W. Sams engineering staff. Indianapolis, Ind.: Howard W. Sams, 1980. 128p. illus. index. ISBN 0-672-21748-1. $6.50.

A guide to more than 6,000 receiving tube and 4,000 picture tube direct substitutes for both color and black and white. Also included are 300 industrial substitutions for receiving tubes and 600 communications substitutes. Includes pinouts. Because tube technology is still present in many types of devices, this book is quite useful.

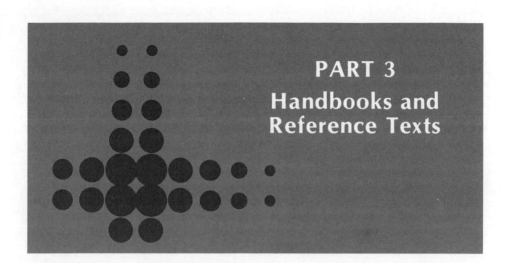

PART 3
Handbooks and
Reference Texts

# 13 Handbooks

Technical handbooks are an important source of information as they are often one of the first places students, librarians, and working engineers look for specific information because they are designed to provide quick access to an enormous variety of basic and sophisticated technical information. Like dictionaries, handbooks are designed to be used on a regular basis. In order to facilitate this regular usage, most handbooks consist of a one-volume compilation of information on specific topics along with related information from other disciplines. They also usually include extensive indexes. As important as organization is to a handbook, the clarity of the text and the accuracy of the data is even more important. Accuracy is imperative because this type of book is often used to check numerical data, definitions, and to find descriptions of processes and materials. Most of the best handbooks give the reference source of the information or data and present lists of additional readings.

Because this section contains a large number of books it has been divided into "Basic Handbooks," "Design," "Fabrication and Manufacturing," "Power and Power Distribution," and "Radio, Television, and Telecommunications."

## Basic Handbooks

203. **Basic Electronic Instrument Handbook.** Edited by Clyde F. Coombs, Jr. New York: McGraw-Hill, 1972. 832p. illus. index. LC 72-1394. ISBN 0-07-012615-1. $54.50.

This heavily illustrated book covers the instruments used in electronics, including oscillators, pulse generators, microwave signal generators, electronic counters, timekeeping instruments, and AC and DC voltmeters. Information is provided on transducers, ammeters, amplifiers, and frequency synthesizers. This basic book was designed to bridge the gap between electronic engineers and instrument users in other scientific fields, like biology. A good discussion is included of what instruments readouts may mean and how errors in interpretation may occur by taking the readout too literally. Coombs is the author of several well-known electronics reference books.

204.   **Electric Motor Handbook.** Edited by E. H. Werninck. London: McGraw-Hill UK Limited, 1978. 629p. illus. index. LC 77-30474. ISBN 0-07-084488-7.

Details all types of motors from direct current and alternating current to special purpose motors and electric controls. The book is directed at anyone who uses, applies, or maintains electric motors. References are provided at the end of some sections. Includes a good basic overview of information and performance characteristics of electric motors.

205.   **Electrical Engineering Handbook.** Compiled by Siemens AG. New York: Wiley Heyden, 1982. 762p. illus. index. ISBN 0-471-26020-7. $87.95.

Based on the German edition, which was compiled by specialists in all branches of electrical engineering from the large German electrical firm of Siemens, this book was updated while being translated and contains many illustrations with formulas, graphs, and tables. Includes fundamentals such as the properties of materials, power engineering, and power engineering applications such as industrial heating, control engineering, and automation. Electronics is given scanty coverage as befits a book written by a major German electrical firm.

206.   **Electrical Engineer's Reference Book.** 14th ed. Edited by M. A. Laughton and M. G. Say. Boston: Butterworth, 1985. various paging. illus. index. ISBN 0-408-00432-0. $99.95.

This long-standing British reference book for electrical engineers was revised and updated for the new edition. New or revised chapters include control system analysis, steam generating, cables, and power system operation among others. Other areas covered included mining, power electronics, alternating current generators, and HVDC transmission.

207.   **Electronic Data Reference Manual.** By Matthew Mandl. Reston, Va.: Reston Publishing, 1979. 323p. illus. index. LC 79-12681. ISBN 0-8359-1641-3. $29.95.

Intended to be used as a reference manual by electronics or electrical engineers, this volume contains basic equations, circuit explanations, tables, graphs, solid-state theory, and design factors. It also includes related data on lasers, holograms, integrated circuitry, digital circuits, modulation types, antennas, symbols, and color codes. The author has written numerous electronics books.

208.   **Electronic Databook.** 3rd ed. Edited by Rudolf F. Graf. Blue Ridge Summit, Pa.: Tab Books, 1983. 416p. illus. index. LC 82-19407. ISBN 0-8306-0138-4. $24.95.

Previously published as the *Electronic Design Data Book*, this handbook makes heavy use of tables, charts, and graphs. Designed to be a source book for electronics designers, it covers the electrical and electronic properties of materials and components, and circuit analysis and design. The organization is a progression from smaller systems to larger systems. Each section is structured to be independent with a separate table of contents and bibliography.

209. **Electronic Engineer's Reference Book.** 5th ed. Edited by F. F. Mazda. Woburn, Mass.: Butterworth Publishers, 1983. various paging. illus. index. ISBN 0-408-00589-0. $62.95.

This volume, which represents an extensive revision and expansion of the fourth edition (Butterworth Publishers, 1976), contains 32 new chapters emphasizing recent developments in semiconductor memories, microprocessors, and filters. Topics that have been excluded or de-emphasized include broadcasting, valve technology, and outdated applications. The 62 chapters were written primarily by British experts and are organized into five sections: mathematical functions; circuit theory and statistics; physical phenomena such as electricity, light, and radiation; materials and components; and electronic design and applications such as communications and recording techniques.

210. **Electronics Data Handbook.** 2nd ed. By Martin Clifford. Blue Ridge Summit, Pa.: Tab Books, 1972. 255p. illus. index. LC 72-82250. ISBN 0-8306-2118-x.

A practical book which is heavy on formulas and the explanatory material necessary to clarify the use of a formula or derivation. Chapters include DC circuits, AC circuits (the largest part of the book), vacuum tubes and vacuum tube circuits, semiconductors, and television antennas and transmission lines. Tables for mathematical symbols and electronic abbreviations are included. The title is a bit misleading since the emphasis is on electrical engineering rather than electronics.

211. **Electronics Designer's Handbook.** 2nd ed. Edited by L. J. Giacoletto. New York: McGraw-Hill, 1977. various paging. illus. index. LC 77-793. ISBN 0-07-023149-4. $86.50.

This handbook covers electronic components, circuit analysis, and circuit design. Each section is structured independently with its own table of contents, footnotes, and bibliography of related material. All sections contains extensive tabular and graphical information. Many detail specific components or systems such as wave-generating circuits, power supplies, receivers, and filters.

212. **Electronics Engineers' Handbook.** 2nd ed. Edited by Donald G. Fink. New York: McGraw-Hill, 1982. various paging. illus. index. LC 74-32456. ISBN 0-07-020981-2. $52.95.

One of the standard handbooks in electronics, this volume is familiarly known as "Fink's." The companion to *Standard Handbook for Electrical Engineers*, the intent of the book is: to compile in one place all pertinent data; to be oriented toward applications; and to provide only enough theory to assure a basic understanding of the concepts included in the book. There are four sections, the first of which covers the principles used in electronics. Section two deals with materials, devices, components, and assemblies, while section 3 deals with electronic circuits and functions. Section 4 treats systems and applications. Over 2,125 illustrations and 3,285 bibliographic entries are included. This is the one handbook that should be found in all electronic collections and should be owned by all electronic engineers.

213.    **Electronics Handbook.** By Clyde N. Herrick. Pacific Palisades, Calif.: Goodyear Publishing Co., 1975. 225p. illus. index. LC 74-15621. ISBN 0-87620-266-0.

Divided into 20 sections including variational characteristics of circuits, digital logic truth tables, and modulated waveform characteristics. The specific topics covered are presented in a graphical format making heavy use of universal charts to save the time of working engineers and technicians. Little coverage is devoted to algebraic analysis. A glossary of semiconductor terms is included.

214.    **Electronics Handbook.** By Matthew Mandl. Reston, Va.: Reston Publishing, 1983. 360p. index. LC 83-3414. ISBN 0-8359-1603-0. $34.95.

A broad range of electronic topics are covered, from basic radio and television to communications and digital circuitry. Intended as a reference text for students, technicians, and engineers, this book is lavishly illustrated with schematics. Chapter 12 includes an extensive listing of letter abbreviations and definitions of commonly used words and phrases in electronics. Definitions provide the section and page number for the discussion or usage of the phrase or word in the text. There is an extensive index with numerous cross-references.

215.    **Electronics Ready Reference Manual.** By Edward Pasahow. New York: McGraw-Hill, 1985. 565p. illus. index. LC 84-11280. ISBN 0-07-048723-5. $28.95.

This pocket-sized reference book is designed to provide a single source for the current electronics information needed by engineers, technicians, students, and hobbyists. The manual contains formulas, tables, and diagrams that provide straight forward solutions to electronic circuits and problems. Coverage includes general laws of electronics, passive and active devices, linear circuits, filters, power supply and regulation, electronic measurement, digital circuits, microprocessors, electronics mathematics, and properties of electronic materials.

216.    **Encyclopedia of Electronic Circuits.** By Rudolf R. Graf. Blue Ridge Summit, Pa.: Tab Books, 1985. 768p. illus. index. LC 84-26772. ISBN 0-8306-1938-0. $50.00.

Over 1,700 circuit designs from 100 different sources are covered, including manufacturers data books, electronic journals, and government publications. A source number is printed on each schematic with the citation being listed in the back of the book. Many of the circuit diagrams include a short explanatory note. The circuits are organized by application, such as alarm, attenuators, crossover networks, modulators, power supplies, and waveform. The schematic drawings are clear, well-labeled, and indexed. This useful book for hobbyists and students can be used as an entry point for manufacturers data books. The compiler is a senior member of the IEEE.

217.    **Encyclopedia of Electronics.** Edited by Stan Gibilisco. Blue Ridge Summit, Pa.: Tab books, 1984. 1,024p. illus. index. LC 84-16437. ISBN 0-8306-2000-1. $60.00.

This encyclopedia, heavily illustrated with graphs, charts, photographs, and drawings, has over 3,000 entries arranged alphabetically. Terms are defined and cross-referenced where relevant. The areas of electronics and electronics applications which are covered include digital electronics theory and applications, microcomputers, radio, cable television, circuit protection, and basic electronic terms and theories. The heavy use of qualitative and quantitative data makes this book a welcome addition to the literature.

218. **Handbook for Electronic Engineering Technicians.** 2nd ed. By Milton Kaufman and Arthur Seidman. New York: McGraw-Hill, 1984. 544p. illus. index. ISBN 0-07-033408-0. $34.95.

An extensive revision of the first edition, this handbook treats analog and digital circuit technology in a practical, nontheoretical manner. Because it is intended to serve as a reference for electronics technicians, it stresses laboratory efficiency, wiring, soldering, and other practical tasks. This is a useful book for all electronics collections.

219. **Handbook of Batteries and Fuel Cells.** Edited by David Linden. New York: McGraw-Hill, 1983. 1,088p. illus. index. LC 82-23999. ISBN 0-07-037874-6. $75.00.

A compilation of the work of 40 experts, this book covers primary batteries, secondary batteries, advanced secondary batteries, reserve batteries, and fuel cells. Extensive use is made of charts, tables, graphs, drawings, and other illustrations. A glossary of terms with definitions and cross-references appears as an appendix. References are provided at the end of most chapters.

220. **Handbook of Computers and Computing.** Edited by Arthur H. Seidman and Ivan Flores. New York: Van Nostrand Reinhold, 1983. 874p. illus. index. LC 83-16942. ISBN 0-442-23121-0. $77.50.

Fifty signed chapters include coverage of components like IC logic families and computer memories; devices like displays and modems; hardware systems like robotics and automatic test equipment; languages like APL, ADA, Assembly, and Fortran; and software systems and procedures like security and installation. Charts, tables, and graphs are included in the text. There are references at the end of most chapters. A large index includes over 3,500 terms. It is differentiated from other handbooks on this subject by the large number of program excerpts.

221. **Handbook of Electronic Circuits and Systems.** By Matthew Mandl. Reston, Va.: Reston Publishing, 1980. 400p. illus. index. LC 80-13298. ISBN 0-8359-2738-5. $29.95.

Produced by a well-known electronics writer, this book is intended as a reference book of electronic fundamentals, applications, and systems. Generously illustrated with schematic drawings, the book covers all areas from measurement and analysis, coils, capacitors to transistors, resonant circuits, and audio and RF power amplifiers.

222. **Handbook of Electronic Formulas, Symbols and Definitions.** 2nd ed. By John R. Brand. New York: Van Nostrand Reinhold, 1979. 368p. index. LC 78-26242. ISBN 0-442-20999-1. $22.95.

This collection of formulas is arranged alphabetically by symbol with the exception of some passive circuit symbols. Formulas may be accessed through the index if the symbol is unknown. Alternate methods, approximations, and schematic diagrams are presented. Each formula uses only the basic units. Divided into "Passive Circuits," "Transistors," and "Operational Amplifiers."

223. **Handbook of Electronic Materials.** Compiled by the Electronic Properties Information Center. New York: IFI/Plenum. 1971-1976. LC 76-147312.

A nine-volume set separately authored and indexed covering:

*Optical Properties.* Vol. 1. By Alfred James Moses. New York: IFI/Plenum. 110p. ISBN 0-306-67101-x. $55.00.

*III-V Semiconducting Compounds.* Vol. 2. By M. Neuberger. New York: IFI/PLenum. 61p. ISBN 0-306-67107-7. $55.00.

*Silicon Nitride for Microelectronic Applications.* Vol. 3. By John Milek. New York: IFI/Plenum. pt 1. 126p. ISBN 0-306-67103-4. $55.00.

*Niobium Alloys and Compounds.* Vol. 4. By M. Neuberger. New York: IFI/Plenum. 70p.

*Group IV Semiconducting Compounds.* Vol. 5. By M. Neuberger. New York: IFI/Plenum. 65p. LC 74-2871. $55.00.

*Silicon Nitride for Microelectronics.* Vol. 6. By John Milek. New York: IFI/Plenum. 124p. pt. II. ISBN 0-306-67106-9. $55.00.

*III-V Ternary Semiconducting Compounds Databook.* Vol. 7. By M. Neuberger. New York: IFI/Plenum. 61p. ISBN 0-306-67107-7. $55.00.

*Linear Electrooptic Modulator Materials.* Vol. 8. By John Milek. New York: IFI/Plenum. 256p. ISBN 0-306-67108-5. $55.00.

*Electronic Properties of Composite Materials.* Vol. 9. By Maurice Leeds. New York: IFI/Plenum. 108p. ISBN 0-306-67109-3. $66.00.

224. **Handbook of Lasers, CRC: With Selected Data on Optical Technology.** Edited by Robert J. Pressley. Cleveland, Ohio: CRC Press, 1971. 631p. index. LC 72-163066. ISBN 0-87819-381-2. $49.95.

Intended for researchers active in lasers, this handbook includes both published and unpublished critically evaluated data, presented primarily as tables. References to the original work containing the data are cited below the relevant tables. Examples of tables include multilayer dielectric coatings, linear electro-optic materials, and holographic parameters and recording materials.

225. **Handbook of Materials and Techniques for Vacuum Devices.** By Walter Kohl. New York: Van Nostrand Reinhold, 1967. 623p. illus. index. LC 67-18288.

Previously published as *Materials and Techniques for Electron Tubes,* this book is of interest primarily for its coverage of microwave tubes, particle accelerators, and thermonic energy converters. Extensive properties information is provided for such materials as glass, ceramics, mica, carbon and graphite, iron and steel, copper and copper alloys, etc. The section on joining processes, especially glass to metal and ceramic to metal, is of particular interest.

226. **Handbook of Modern Electronics Math.** By Sam Cowan. Englewood Cliffs, N.J.: Prentice-Hall, 1983. 254p. index. LC 82-11260. ISBN 0-13-380485-2. $21.95.

A ready reference for mathematical formulas and methods used in solving electronic problems, this book includes nomograms for determining inductive and capacitive reactance; Norton's and Thevenem's theorems; discussions on how to convert decimal, binary, and hexidecimal numbers; equations for resistor values; and three methods for simplifying logic circuits.

227. **Handbook of Rotating Electric Machinery.** By Donald V. Richardson. Reston, Va.: Reston Publishing, 1980. 652p. illus. index. LC 78-926216. ISBN 0-8359-2759-8. $26.95.

Specifically designed for engineers who have never actually worked with electric motors, this volume's goal is to help the user recognize, understand, analyze, specify, connect, control, and satisfactorily apply the various types of electric motors and generators. In the past it was common for departments of electrical engineering to teach a course on electric motors and machinery. This is no longer the case and, as a result, many engineers do not have experience with electrical machinery. This book is valuable in filling this gap.

228. **Handbook of Semiconductor Electronics: A Practical Manual Covering the Physics, Technology, and Applications of Transistors, Diodes and Other Semiconductor Defects in Conventional and Integrated Circuits.** 3rd ed. Edited by Lloyd R. Hunter. New York: McGraw-Hill, 1970. 2v. illus. index. ISBN 0-07-031305-9.

An outstanding handbook covering the field of semiconductor electronics. Each section is developed in typical handbook format with historical and general considerations presented first, followed by the specifics of the topic. The book is divided into four parts: "Physics of Semiconductor Materials," "Devices and Circuits," "Technology of Semiconductor Devices," and "Measurement and Analytical Techniques."

229. **Handbook on Semiconductors.** Edited by T. S. Moss. New York: Elsevier Science Publishing, 1982. 4v. illus. index. LC 79-16500. ISBN 0-444-85298-0. $631.75/set.

These comprehensive volumes covers all aspects of semiconductors. Each section was written by an expert in the field. References are included at the end of each section, many of which are from the early 1980s.

Volume 1 details basic crystal structure, energy bands, and transport properties like scattering phenomena, and quantum and polaronic transport. The intrinsic and extrinsic optical properties of solids are covered in volume 2, including effects of temperature, pressure, electric field, and doping. Methods of preparation including crystal growth, purification, doping, and implanting are treated in volume 3. Characterization is also covered in depth. Volume 4 treats the physics of many types of semiconductor devices including different diode types, bipolar transistors, and various field effect devices. These books should be in any collection which includes solid-state physics or electronics materials.

230. **How to Measure Anything with Electronic Instruments.** By John A. Kuecken. Blue Ridge Summit, Pa.: Tab Books, 1981. 336p. index. LC 81-9134. ISBN 0-8306-0025-6. $21.95.

Shows measurement techniques using electronic devices and describes methods for creating instruments which are easy to use and give accurate readings. A good book for hobbyists, technicians, and students interested in electronic projects, it also is a good book for public and community college library collections.

231. **IES Lighting Handbook 1984.** rev. ed. New York: IES Publication Sales, 1984. 2v. illus. index. LC 84-19818. ISBN 0-87995-007-2. $107.00/set.

This standard handbook was divided into two parts in 1981 so that it could be updated more quickly. Volume 1, the reference volume, covers color, daylighting, light sources, lighting calculations, and the fundamental physics of light. Volume 2 deals with applications of lighting including energy management, residential lighting, and institution and public building lighting. This last category covers banks, libraries, and museums. Chapter 1 of volume 1 consists of a "Dictionary of Lighting." An index to both volumes is found in volume 2. This set should be on the reference shelf of most libraries.

**232.  Illustrated Handbook of Electronic Tables, Symbols, Measurements and Values.** 2nd ed. Edited by Raymond H. Ludwig. Englewood Cliffs, N.J.: Prentice-Hall, 1984. 415p. illus. index. LC 83-17620. ISBN 0-13-450494-1. $27.95.

Concentrates on standardized electronic data, including symbols, abbreviations, and related reference data. A pictorial directory of electronic symbols is included. Of special interest are the chapters on key facts and formulas, which are illustrated with solved problems. A broad range of ready reference, hard-to-find, timesaving information like wire gage, drill and tap sizes, machine screw sizes, and radio and television frequency allocations can be found in this book.

**233.  Infrared Handbook.** William L. Wolfe and George J. Zissis. Washington, D.C.: Office of Naval Research, 1978. various paging. index. LC 77-90786. ISBN 0-9603590-0-1. $25.00.

With extensive references and bibliographies, this work is of interest primarily for its sections on charge coupled devices, imaging tubes, displays, and detector associated electronics.

**234.  Master Handbook of Electronic Tables and Formulas.** 4th ed. Edited by Martin Clifford. Blue Ridge Summit, Pa.: Tab Books, 1984. 392. diagrams. LC 80-14358. ISBN 0-8306-0625-4. $21.95.

A compilation of electronic data arranged in tabular form. In most cases the answer is supplied without the need for much mathematics. The areas covered include voltage and current, capacitance, sound and acoustics, conversions, time constants, thermocouples, and digital logic. Designed for use by technicians and engineers who do not work in electronics regularly, the clear presentation makes this a useful book for students.

**235.  Standard Handbook for Electrical Engineers.** 12th ed. Edited by Donald Fink and H. Wayne Beatty. New York: McGraw-Hill, 1987. various paging. illus. index. LC 56-6964. ISBN 0-07-020975-8. $85.50.

This is probably one of the best, most well-known electrical handbooks. Now in its 12th edition, it has been used by generations of engineers and librarians and is referred to familiarly as "finks." One of the best features of this book is its heavy use of tables, graphs and charts to present data. Major sections include wiring design, motors, illumination, electricity in transportation, telecommunications, and electrotechnology. Substantial coverage is given to generation, transmission, distribution, and control of electric power. Information on nuclear power, alternative energy sources, and the applications of computer technology and electronics were added as a major revision. This book should be in all electrical collections and should be owned by all electrical engineers and students. The editor is a well-known author of electrical reference materials.

236. **Standard Handbook of Engineering Calculations.** 2nd ed. Edited by Tyler Gregory Hicks. New York: McGraw-Hill, 1985. various paging. illus. index. LC 84-28929. ISBN 0-07-028735-x. $59.50.

Provides specific step-by-step calculation procedures for the most common design and operating problems. Sections 4 and 5 cover electrical engineering and electronics. Examples of calculations include direct-current circuit analysis, Krichhoff's Laws for DC circuit analysis, power system short circuit current, power factor determination, buffer amplifier analysis, low and high pass filter design, microwave transmitter analysis, voltage, and current gain. Each section provides references, diagrams, charts, and tables along with the calculations.

# Design

This category includes handbooks which contain information of interest to engineers designing new integrated circuits or new devices using "off the shelf" components.

237. **Amplifier Handbook.** Edited by Richard R. Shen. New York: McGraw-Hill, 1966. various paging, illus. index. LC 64-66296.

Although this book is now out of print, the format and amount of material covered make this book enormously popular with students. The characteristics, properties, and applications of each type of amplifier are clearly spelled out in the 50 signed chapters.

238. **Communications Circuits Ready Reference.** By John Markus. New York: McGraw-Hill, 1982. 160p. illus. index. LC 82-218. ISBN 0-07-040460-7. $12.50.

Like the other Markus's ready reference guides in format and data sources, the organization of this book is by families of ICs with each chapter covering a different family, including operating amplifiers, functional circuits, design of active filters using operational amplifiers, phased lock loops, timing circuits, and IC power management circuits. The emphasis is on the most useful and cost-effective solution to common design problems. Clear circuit diagrams are provided.

239. **The Complete Handbook of Amplifiers, Oscillators and Multivibrators.** By Joseph J. Carr. Blue Ridge Summit, Pa.: Tab Books, 1981. 364p. illus. index. LC 80-28375. ISBN 0-8306-1230-0. $10.95.

By the same author who wrote *The Complete Handbook of Radio Transmitters* and *The Complete Handbook of Radio Receivers*, this basic book covers the theory, design, and use of the basic building blocks of electronics, amplifiers, oscillators, and multivibrators. It is designed for all levels of user, from hobbyist to designer.

240. **Designer's Handbook of Integrated Circuits.** By Arthur D. Williams. New York: McGraw-Hill, 1984. various paging. illus. index. LC 82-14955. ISBN 0-07-070435-x. $28.95.

Organized by application rather than by IC type, each section gives fundamentals and theory along with application and section information. The author provides fairly extensive comparisons between ICs by different manufacturers. All major applications are covered including operation amplifiers, telecommunication circuits, phase-lock loops, SSI and MSI logic circuits, and optoelectronics. References at the end of each section are not limited to books and articles, but includes manufacturers data sheets which add value to the book.

241. **Directory of Electronic Circuits: With a Glossary of Terms.** revised and enlarged. By Matthew Mandl. Englewood Cliffs, N.J.: Prentice-Hall, 1978. 321p. illus. index. LC 78-30879. ISBN 0-13-214924-9.

Designed as a reference book to all electronic circuits, this directory includes control, communication, signal processing, and digital circuits. It is divided into 15 chapters containing a total coverage of 200 circuits. In addition to circuit diagrams, basic formulas, and data, there is a glossary of common electronic terms. Most of the circuits are adequately described, but some subjects, such as digital circuits, are not covered in sufficient depth for use by the professional circuit designer.

242. **Electronic Components Handbook.** By Thomas H. Jones. Reston, Va.: Reston Publishing, 1978. 391p. illus. index. LC 77-22341. ISBN 0-87909-222-x. $25.95.

Covers in detail fixed capacitors; variable capacitors; transformers; inductors; fixed, variable, and nonlinear resistors; relays; etc. Performance characteristics are provided for industrial varieties, military, and hobbyist components. Only components widely available either as stock items or through mail order are included. The list of relevant data books from electronics companies is a unique feature. Each chapter ends with a bibliography and definitions of terms.

243. **Electronic Filter Design Handbook.** By Arthur B. Williams. New York: McGraw-Hill, 1981. illus. index. LC 80-11998. ISBN 0-07-070430-9. $54.50.

This handbook treats the design of active and passive filters in a practical manner helping the average engineer without previous experience design almost any type of filter. Coverage includes selecting response characteristics, low pass, high pass, bandpass filters, design of magnetic components, and component selection for active filters. Very popular with engineering students.

244. **Encyclopedia of Electronic Circuits.** By Rudolf F. Graf. Blue Ridge Summit, Pa.: Tab Books, 1985. 760p. illus. LC 84-26772. ISBN 0-8306-0938-5. $29.95.

This encyclopedia provides approximately 1,300 schematic illustrations of analog and digital circuits which have been excerpted from major electronics publications. The circuits are divided into 98 basic categories and include schematics for instruments and appliances. Each schematic is numbered and keyed back to the original source document.

245. **Encyclopedia of Integrated Circuits: A Handbook of Essential Reference Data.** By Walter H. Buchsbaum. Englewood Cliffs, N.J.: Prentice-Hall, 1981. 384p. illus. index. LC 80-21596. ISBN 0-13-275875-k. $24.95.

Arrangement is by type of circuit: analog, consumer, digital, etc. Because this handbook attempts to cover the entire field of ICs from the point of view of the users rather than from that of the designers, it is not concerned with part numbers or interchangeability. Coverage is by the ICs function and how it performs. Each entry includes a description, list of key parameters with explanation, and applications data.

246. **Guidebook of Electronic Circuits.** By John Markus. New York: McGraw-Hill, 1974. 992p. illus. index. LC 74-9616. ISBN 0-07-040445-3. $75.00.

Covers more than 3,600 modern circuits and circuit applications. For each circuit, values for all important components, concise descriptions, performance data, and suggested applications are given. Among the 131 applications covered are burglar alarm, motor control, regulator, telephone circuit, and zero voltages. This is a good example of how this type of reference book should be organized and the type of information that should be given. While a bit dated, it still gives a large amount of data and information on circuits that the user might find in existing applications. Citations to the original source are listed for each circuit.

247.  **Handbook of Electronic Circuits.** By John D. Lenk. Englewood Cliffs, N.J.: Prentice-Hall, 1976. 307p. illus. index. LC 75-8635. ISBN 0-13-377309-4.

Like Lenk's other books, this one emphasizes the simple, practical approach while covering a wide range of circuit designs. Included are filters, attenuators, wave generators, phototransistor circuits, switching, and oscillator circuits.

248.  **Handbook of Electronic Circuits and Systems.** By Matthew Mandl. Reston, Va.: Reston Publishing, 1980. 400p. illus. index. LC 80-13298. ISBN 0-8359-2738-5. $29.95.

Intended as a reference of electronic fundamentals, applications, and systems, this book is heavily illustrated with schematic drawings. It covers all areas from measurement and analysis, coils, and capacitors to transistors, resonant circuits, and audio and RF power amplifiers.

249.  **Handbook of Electronic Systems Design.** By Charles A. Harper. New York: McGraw-Hill, 1980. various paging. illus. index. LC 79-19103. ISBN 0-07-026683-2. $57.50.

Deals with general purpose computers including guidelines for the design of both components and subsystems. Also covered are applications and usages of communication networks and systems. Treated in some detail are sinks, signal processing, wirelines, and satellites. An entire chapter is devoted to radar systems and subsystems.

250.  **Handbook of Electronics Packaging Design and Engineering.** By Bernard Matisoff. New York: Van Nostrand Reinhold, 1982. 400p. illus. index. LC 81-2362. ISBN 0442-20171-0. $37.50.

Designed to be a reference book for frequently encountered design problems including structures, mechanisms, materials, finishes, appearance, utility, serviceability, and reliability. Particular emphasis is placed on answering questions that relate to heat transfer, electrical design, industrial design, and techniques to ease assembly procedures. Interchangeability of parts, consumer appeal, and ergonomic considerations are covered in some detail.

251.  **Handbook of Integrated Circuits: For Engineers and Technicians.** By John D. Lenk. Reston, Va.: Reston Publishing, 1978. 480p. illus. index. LC 78-7260. ISBN 0-8359-2744-x. $34.95.

A companion to the author's *Manual for Integrated Circuit Users,* this book is written for users rather than designers. As a result only existing commercial ICs are covered. Areas included are op amps, linear packages and arrays, and practical considerations like mounting, soldering, powering, and heat sinks.

252.  **Handbook of Modern Solid-State Amplifiers.** By John Lenk. Englewood Cliffs, N.J.: Prentice-Hall, 1974. 414p. illus. index. LC 73-21834. ISBN 0-13-380394-5. $25.95.

All types of amplifiers in current usage are detailed, including audio, radio frequency, direct coupled, differential, and compound and operation amplifiers. Also covers discrete amplifier circuits, FETs, and selected ICs.

253.  **Illustrated Encyclopedia of Solid-State Circuits and Applications.** By Donald R. Mackenroth and Leo G. Sands. Englewood Cliffs, N.J.: Prentice-Hall, 1984. 353p. illus. index. LC 83-23077. ISBN 0-13-450537-9. $29.95.

Heavily illustrated with schematics, this book provides an explanation of the workings and characteristics of each semiconductor and basic circuit. The aim of this handbook is to develop a clear understanding of the principles and applications of solid-state circuitry so that the user can specify and use these devices.

254.  **Integrated Circuits Applications Handbook.** Edited by Arthur H. Seidman. New York: John Wiley & Sons, 1983. 673p. index. LC 82-10903. ISBN 0-471-07765-8. $39.95. (Wiley Electrical and Electronics Handbook Series).

This very popular overview/handbook includes many definitions, schematics, graphs, and tables. The major emphasis is on IC applications. Three major areas covered are digital ICs, including TTL logic and counters; linear ICs, including op amp and phase lock loops; and fabrication using both thick and thin film techniques.

255.  **Master Op Amp Applications Handbook.** Edited by Harry W. Fox, Jr. Blue Ridge Summit, Pa.: Tab Books, 1977. 467p. illus. index. LC 77-20844. ISBN 0-8306-6856-x. $13.95.

Covers IC circuits, ideal and nonideal IC op amps; basic op amps circuits with audio applications; converting signals with IC op amps; instrumentation and transducer amplifiers; integrators; active filters; nonlinear circuits using op amps, pulse, digital, and switching applications; and wave form generators. Good coverage of this important area.

256.  **Materials and Processes in Electronics.** By C. E. Jowett. London: Hutchinson and Co., 1982. 328p. index. LC 81-165691. ISBN 0-09-45100-0.

This book was written to help the design engineer select materials for components, structures, and mechanisms. Both the properties of different materials and the methods used in applying those materials are covered. Information is given on adhesive bonding, connector materials, metallized ceramic materials, cable lacing, and wafer exposure. This is one of the best books available in this very important area of materials selection.

257.  **Microprocessor and Microcomputer Data Digest.** By W. H. Buchsbaum and Gina Weissenberg. New York: Reston Publishing, 1983. 336p. illus. index. LC 82-21620. ISBN 0-8359-4381-x. $29.95.

Designed to provide basic technical data for each microprocessor IC that is currently available "off-the-shelf." Included are microcomputers which are treated as microprocessors, slide microprocessors, and ICs that contain only a portion of a computer central processing unit (CPU). Each of the different categories is covered in numerical sequence with each entry containing a pin connection diagram, a description of each pin connection, and a brief functional description of the microprocessor. The data provided are "digested" from technical data provided by the manufacturers. Buchsbaum is a well-known electronics writer.

258. **Practical Guide to Digital Integrated Circuits.** 2nd ed. By Alfred W. Barber. Englewood Cliffs, N.J.: Prentice-Hall, 1984. 272p. illus. index. LC 83-21208. ISBN 0-13-690751-2. $21.95.

A helpful, realistic guide to digital ICs which covers the relationship between ICs and transistors. Includes explanations of how circuits have been used in logical ICs, information on how to combine gates and flip-flops, the use of breadboards, and the design of complex systems using the basic elements of electronics. Also covered are ways of analysing large scale integration and tips on selecting and using ICs, including ways of testing ICs.

259. **Sourcebook of Modern Transistor Circuits.** By Laurence G. Cowles. Englewood Cliffs, N.J.: Prentice-Hall, 1976. 360p. illus. index. LC 75-30748. ISBN 0-13-823419-1.

Describes the amplifiers, diodes, and switching circuits used most often by designers. The circuits, shown with component values, are made with easily obtainable semiconductors. Many tables, charts, and schematics are provided. Among the areas covered are two-stage transistors, amplifiers, power supplies, regulators, filters, diodes, circuits, and applications. A short annotated bibliography is included.

260. **User's Guide to Selecting Electronic Components.** By Gerald L. Ginsberg. New York: Wiley-Interscience Publication, 1981. 249p. illus. index. LC 80-25197. ISBN 0-471-08308-9.

Arrangement is by type of component, such as resistors, capacitors, electromagnetic components, power sources, special function components, and solid-state devices. Intended to help engineers select components for equipment being designed, each section contains information about the groups of components and about selected specific components within each group. Relevant terms are defined and applications are given at the end of each section. A bibliography is included.

# Fabrication and Manufacturing

Books in this category contain information that can be used to turn designs into working circuits, devices, and components.

261. **Designing and Creating Printed Circuits.** By Walter Sikonowiz. New York: Hayden Book Co., 1981. 164p. illus. index. LC 81-6524. ISBN 0-8104-0964-x. $8.95.

The intent of this book is to explain and examine the manufacturing process for printed circuits. Good illustrations of the layers of circuits and art work are provided. Of particular interest are the sections covering laminate construction, copper foil, trace routing, automated design methods, photoresist versus screened resist, applications of plated metal, and etching and cleaning. Includes a bibliography.

262. **Electronic Assembly.** By Jeremy Ryan. Reston, Va.: Reston Publishing, 1979. 168p. illus. index. LC 79-17953. ISBN 0-8359-1638-1. $16.95.

Provides for students and technicians a clear logical background of techniques used in electronic assembly, including parts identification, color codes, hand tools, hardware, parts mounting, soldering, harnessing, splicing conductors, wire wrapping, and joining techniques. Of particular interest is the section on harnessing which includes construction techniques for harnesses, lacing the cable, sleeving, and straps and clips.

263. **Handbook of Cabling and Interconnecting for Electronics.** Edited by Charles Harper. New York: McGraw-Hill, 1972. various paging. illus. index. ISBN 0-07-026674-3.

Like other McGraw-Hill handbooks, this text is liberally illustrated with tables and graphs and has references at the end of most sections. Each chapter is written by an expert in the field. Coverage includes soldered, welded, and mechanical terminating systems; basic connector systems; coaxial cables; magnetic wire; rigid printed wiring; and other practical aspects of wiring and cabling used in electrical engineering and electronics.

264. **Handbook of Electrical and Electronic Insulating Materials.** By W. Tillar Shugg. New York: Van Nostrand Reinhold, 1986. 640p. illus. index. LC 86-4104. ISBN 0-317-45618-0. $89.50.

This handbook details 15 classes of commonly used insulating materials, such as plastics, magnetic enamels, and film. Information is provided on manufacturing methods, testing, market trends, and properties. Properties information includes chemistry, grades, processing, standards, and applications.

265. **Handbook of Electronic Packaging.** Edited by Charles A. Harper. New York: McGraw-Hill, 1969. various paging. illus. index. LC 68-11235. ISBN 0-07-026671-9. $56.75.

Electronic packaging consists of overlapping disciplines because it is the conversion of electronic or electrical functions into producable, electromechanical assemblies or packages. The handbook includes practical and useful data and guidelines on materials, components, processes, connection and interconnection techniques and devices, mechanical layouts, electrical factors, and thermal design of both microelectronic and conventional packaging. The individual sections, which were written by experts, include a glossary of terms and references.

266. **Handbook of Electronics Industry Cost Estimating Data.** By T. Taylor. New York: John Wiley, 1985. 464p. illus. index. LC 85-6294. ISBN 0-471-82264-7. $54.95.

A collection of time standards and manufacturing methods that can be used for cost-estimating electronic equipment and systems. Covers machining, sheet metal fabrication, wiring, circuit board assembly, electrical testing, and packaging. Also details selected management and supervisory areas like production planning, scheduling, and personnel ratios.

267. **Handbook of Electronics Manufacturing Engineering.** 2nd ed. By Bernard Matisoff. New York: Van Nostrand Reinhold, 1986. 400p. illus. index. LC 78-9689. ISBN 0-442-25146-7. $44.95.

Nearly one-third of the book is made up of reference tables and a glossary of terms used in metallurgy and fabrication. Tables include fastener head styles, twist drill data, and transistor pin identification. The book is richly illustrated with examples of electronic manufacturing processes. Particularly interesting are the examples of both good and bad manufacturing practices. Pictorial examples of good and bad soldering, wiring, and wrapping are also provided.

268. **Handbook of Printed Circuit Design, Manufacture, Components & Assembly.** By Giovanni Leonida. London: Electrochemical Publications, 1980. 700p. illus. index. LC 82-218. ISBN 901150-08-8.

Aims to provide in a single book the information necessary for the manufacture of printed circuits. Because electronic assembly is not an independent discipline, the book covers, along with electronics, aspects of the following related areas: mechanics, thermodynamics, fluid dynamics, chemistry, metallurgy, and optics. Each of the 10 chapters, which covers a specific topic like components or design, is divided into specific sections like capacitors, monolithic integrated circuits, etc. Heavily illustrated with drawings and photographs and a large number of tables.

269. **Handbook of Printed Circuit Manufacturing.** By Raymond H. Clark. New York: Van Nostrand Reinhold, 1985. 820p. illus. index. LC 84-13117. ISBN 0-442-21610-6. $45.95.

The purpose of this handbook is to acquaint readers with the steps required to manufacture printed circuits, which of all electronic components probably require the most skilled manufacturing operations. This book is a source of hands-on manufacturing know-how, step-by-step procedures, trouble-shooting information, and guidelines explaining each process and technique in detail. Sections include "Design and Manufacture," "Planning, Document Control," "Imaging and Artwork," "N/C Processing," "Plating," "Multilayer Printed Circuits," and "Process Control." The appendixes include soldermasking, printed circuit baths, IR fusing process, and screen printing. Each of the techniques is documented with high-quality photographs.

270. **Handbook of Thick Film and Hybrid Microelectronics.** Edited by Charles A. Harper. New York: McGraw-Hill, 1974. various paging. illus. index. LC 74-2460. ISBN 0-07-026680-8. $65.00.

Covers circuit, design, photofabrication, manufacturing operations, substrates for thick film circuits, conductor and resistor materials, dielectric materials, component attachment techniques, and packaging and interconnection of assemblies. Consists of over 200 tables and figures in signed chapters with references at the end. Generously illustrated with a glossary of terms at the end.

271. **Handbook of Thick Film Technology.** Edited by P. J. Holmes and R. G. Loasby. London: Electrochemical Publications, 1980. 430p. index. illus. ISBN 0-901150-05-3. $145.00.

Provides coverage of development of thick film technology, deposition processes and equipment, screen printing materials and procedures, substrates, conductor materials, characteristics of dielectrics, capacitors, crossovers, and overglazes. Also discusses the structure of resistor compositions, characteristics of resistors, and special purpose materials and processes. Also includes treatment of the following subjects which have direct application to electronics: thick films at high frequencies, thermal design aspects, adjustment of thick film components and circuits, circuit devices and assembly, packaging of hybrid microcircuits, and reliability. Bibliographies are found at the end of most chapters.

272. **Microelectronics Interconnection and Packaging.** Edited by Jerry Lyman. New York: McGraw-Hill, 1980. 320p. illus. index. LC 79-21990. ISBN 0-07-19184-0. $35.00.

A compilation of articles from *Electronics* magazine covering microelectronics interconnections and packaging. Topics include lithography and processing for integrated circuits, thick and thin film hybrids, printed circuit board technology, automatic wiring, IC packages and connectors, environmental factors affecting productions, computer aided design, and automatic testing.

273. **Printed Circuit Boards.** By Neal A. Willison. New York: John Wiley, 1983. 146p. illus. index. LC 82-13541. ISBN 0-471-86177-4. $15.95.

This manual provides step-by-step instruction on how a printed circuit board is made, including the mounting of the electronic components, how the components and boards fit together, as well as assembly, subroutines, encapsulations, and repair. Key technical words are defined and performance objectives are given for each manufacturing step.

274. **Printed Circuits Handbook.** 2nd ed. Edited by Clyde F. Coombs. New York: McGraw-Hill, 1979. various paging. illus. index. LC 78-16800. ISBN 0-07-012608-9. $46.50.

The printed circuit is the major interconnection technique used in the design and manufacture of electronic components and its basic building blocks—plating, etching, metal cladding, plastic machining, and photopolymers—are covered in this heavily revised and updated second edition. Many new fabrication techniques, like dry film resist, are covered. Like the other McGraw-Hill handbooks each chapter was written by an expert. References to relevant industrial standards are included along with a glossary of common terms.

275. **Thick Film Technology and Chip Joining.** By Lewis F. Miller. New York: Gordon and Breach Science Publishing, 1972. 228p. illus. index. LC 79-175344. ISBN 0-677-03440-7. $44.25.

Describes in detail properties of thick films and processes used in manufacture of both the films and components made using the films. Areas covered include screening and paste transfer, thin film conductors, silver palladium electronics, ternary alloy electrodes, glass resistors, chip joining techniques, and power interconnections. A bibliography is included at the end of the text.

276. **Thin Film Technology.** By Robert W. Berry, et al. New York: Van Nostrand Reinhold, 1968. 706p. illus. index. LC 68-25817. ISBN 0-88275-744-x. $37.50.

Written to provide the scientific basis for the methods and materials used in thin film electronics, this book is divided into three sections which cover the methods of film formation, properties of materials as they relate to electronic uses of thin films, and information necessary for the actual application of thin films to integrated circuits.

277. **VLSI Fabrication Principles.** By Sorab K. Ghandi. New York: Wiley-Interscience Publication, 1983. 665p. illus. index. LC 82-10842. ISBN 0-471-86833-7. $52.50.

Discusses the basic principle underlying the fabrication of semiconductor devices and integrated circuits. While the emphasis is on the processes that are useful for VLSI (very large scale integration), many of these same processes are used in the fabrication of discrete devices and integrated circuits. Each chapter discusses principles common to all semiconductors, with separate sections used to discuss some of the problems of VLSI. This is a liberally illustrated reference book. Areas of particular interest are ion implantation, deposited films, etching and cleaning, lithographic processes, and device and circuit fabrication. This is one of the best books on this very important and growing area of technology.

# Power and Power Distribution

Today power and power distribution is taken for granted by most people until the system breaks down and users experience blackouts or brownouts. However, this has not always been the case. In the 1930s and 1940s it was a major domestic policy in the United States to extend power to most regions and large programs like REA (Rural Electrification Association) and TVA (Tennessee Valley Authority) were set up to extend the benefits of power to remote areas. The following consists of handbooks which deal directly with power transmission, components, and power use. While the actual production of power through steam turbines, nuclear power stations, or hydro-stations is not specifically covered, it is included in some of the following books.

278. **American Electricians' Handbook.** 10th ed. New York: McGraw-Hill, 1980. various paging. illus. index. LC 80-14757. ISBN 0-07-013931-8. $49.50.

The aim of this handbook is to provide a compilation of information on electrical equipment and materials. This edition is in accordance with the *1978 National Electric Code.* Coverage includes properties and splicing of conductors, general electrical equipment, batteries, transformers, and electronic and solid-state devices and circuits. Of special interest is the coverage of both exterior and interior wiring and the inclusion of wiring design tables. Most of the data are presented in tabular form. There is very little reliance on the use of higher mathematics in the material.

279. **Electric Cables Handbook.** Edited by D. McAllister. Surrey, Great Britain: Granada, 1982. 896p. illus. index. ISBN 0-246-11467-3. $120.00.

Includes all types of insulated cables used for supplying 100v to 525Kv AC or DC electrical power. Excluded are telecommunications and specialized applications functions. Because transmission cable practices are fairly similar throughout the world the British orientation of this book does not detract from its usefulness. Appendixes include conductor data, mineral insulated wiring cables, PVC, insulated cables, minimum installation bending radii, and a bibliography.

280. **Electric Power System Components, Transformers and Rotating Machines.** By Robert Stein and William T. Hunt, Jr. New York: Van Nostrand Reinhold, 1979. 477p. illus. index. LC 78-18398. ISBN 0-442-17811-2. $26.50.

The basics of todays electric power system are detailed, including Faradays's induction law, magnetic circuits, sinusoidal steady state, transformers, electromechanical energy conversion, synchronous generators and motors, servomotors, etc. Provides details on how a modern electric power system generates, transmits, and distributes power to users.

281. **Electric Utility Systems and Practices.** 4th ed. Prepared by General Electric Co. New York: John Wiley & Sons, 1983. 336p. illus. index. LC 83-3640. ISBN 0-471-04890-9. $45.95.

Describes the design and operation of power system components as well as how these components are integrated into the power system. Each component is described in detail. Areas covered include power transmission, transformers, switchgears, substations, distribution, and system operations. Also included are the different ways of generating power from the old technologies of steam, water, and combustion turbines to the newer technology using nuclear fuel to generate power. References are provided at the end of each section.

282.   **Electrical Characteristics of Transmission Lines.** By Wolfgang Hilberg. Dedham, Mass.: Artech House, 1979. 184p. illus. LC 78-23940. ISBN 0-89006-081-9. $31.00.

An introduction to the calculation of characteristic impedances and specific capacity and inductance of homogeneous cylindrical and conical electrical transmission lines. The book is divided into four sections, with the first section showing how the tabulated formulas in the second section were obtained. The third section consists of diagrams of transmission lines, while the fourth is an extensive bibliography of references on transmission line calculations.

283.   **Electrical Distribution Engineering.** By Anthony J. Pansini. New York: McGraw-Hill, 1983. 434p. illus. index. LC 82-20814. ISBN 0-07-048454-6. $28.95.

The major aspects of power distribution in the United States is described along with an overview of the systems development. Of particular interest are the sections on load characteristics, design of overhead and underground components and systems, and discussion of metering systems. Also included are the materials and components used including conductors, poles, cross arms, pins, racks, and insulators.

284.   **Electrical Wiring Handbook.** By Edward L. Safford. Blue Ridge Summit, Pa.: Tab Books, 1980. 432p. illus. index. ISBN 0-8306-9932-5. $17.95.

Provides information on several types of wiring including circuits and electrical systems layout, lighting, and electrical renovations. Also has examples of wiring systems, such as the techniques used in wiring security systems. Intended for electricians, supervisors, architects, and engineers, this is a useful book since most architects and engineers have had only a cursory introduction to electrical wiring systems.

285.   **Handbook of Electric Power Calculations.** Edited by Arthur H. Seidman, et al. New York: McGraw-Hill, 1983. various paging. illus. index. LC 82-24910. ISBN 0-070-56061-7. $39.50.

Contains detailed, step-by-step procedures for calculating almost 300 commonly occurring problems in electrical engineering, with an emphasis on practical problem solving. Divided into 20 sections, it includes network analysis, DC and AC machines, transformers, transmission lines, system stability, and grounding.

286.   **Handbook of Fiber Optics: Theory and Applications**. Edited by Helmut F. Wolf. New York: Garland Publishing, 1980. 560p. index. LC 78-31977. ISBN 0-8240-7054-2. $62.50.

All aspects of fiber optics are covered, including optical waveguides, detectors, optical sources and connectors, couplers and switches, optical component applications, and optical communications activities. The last section, written by the editor, is an overview of activities and systems in the United States, Europe, and Japan. Extensive references are provided at the end of each section, while the book ends with a separate author and subject index.

287.   **Handbook of Power Generation: Transformers and Generators.** By John E. Traister. Englewood Cliffs, N.J.: Prentice-Hall, 1982. 272p. illus. index. LC 82-20480. ISBN 0-13-380816-5. $21.95.

Designed to show available equipment, and how to use this equipment for specific applications. Only limited coverage is given to theories. Among the areas covered are principles and characteristics of AC generators, transformer construction, operations, connections, parallel operations, pole and platform mounting, underground wiring, and grounding. A glossary of terms is included.

288. **The J and P Transformer Book, Being a Practical Technology of the Power Transformer.** 11th ed. By A. C. Franklin, D. P. Franklin, and S. Austin Stigant. London: Butterworth Publishers, 1983. 770p. illus. index. LC 83-10162. ISBN 0-408-004-940. $25.00.

This venerable British handbook was first published in 1924. The opening chapters cover fundamental principles, magnetics, transformer efficiencies, connections, and transformer tappings. Several appendixes as well as a subject index are included. Provides a comprehensive discussion of transformers in SI units along with incorporating the latest British and international standards.

289. **Lineman's and Cableman's Handbook.** 7th ed. By Edwin B. Kurtz and Thomas M. Shoemaker. New York: McGraw-Hill, 1986. various paging. illus. index. LC 85-17120. ISBN 0-07-035686-6. $54.50.

A heavily illustrated handbook intended for linemen, cablemen, foremen, and supervisors involved in transmission and distribution of power. The basics of electricity and electric power are discussed, but most of the book is devoted to information about the actual construction of overhead and underground distribution and transmission lines. Also covered are substations, lightning protection, transmission circuits, pole structures, grounding line conductors, and voltage regulation. This is the only book which deals with this specialized topic.

290. **Microwave Engineering.** By Arthur Harvey. New York: Academic Press, 1963. 1,313p. illus. index. LC 62-13090.

While this book is out of print and somewhat dated, it is crammed with useful information and is still popular with users. Each section has extensive references. In fact, the chapter on receivers has 422 entries from the early 1900s-1962. Coverage falls into three interrelated groups: passive components, electrical behavior of microwave systems and devices, and techniques associated with microwave applications.

291. **Microwave Engineers Handbook.** Edited by Theodore Saad. Dedham, Mass.: Artech House, 1971. 2v. illus. LC 76-168891. ISBN 0-89006-004-5, 0-89006-002-9. $43.00/set.

Both volumes include charts, graphs, and tables of data on directional couplers, antennas, ferrites, detection and noise, and microwave tubes. Most of the data are from other handbooks and company literature. The lack of an index hampers the usefulness of this book.

292. **Modern Power Station Practice.** 2nd ed. By Central Electricity Generating Board. Oxford: Pergamon Press, 1971. 8v. illus. LC 71-86200. ISBN 0-08016436-6. $180.00/set.

The contents of the eight volumes are as follows: *Planning and Layout, Mechanical-Boilers: Fuel and Ash Handling Plant, Mechanical Turbines, Electrical-Generators and Electrical Plant, Chemistry and Metallurgy, Instrumentation, Controls and Testing, Operations and Efficiency,* and *Nuclear Power Generation.* Of special interest is the extensive use of diagrams for plant layout, switchgears, toroidal headers, ash slurry pumps, etc. However, the lack of a central index and the British orientation of the book do hamper usefulness.

293.  **Power Generation Calculations Reference Guide.** Edited by Tyler G. Hicks. New York: McGraw-Hill, 1987. 384p. ISBN 0-07-028800-3. $36.50.

Provides calculation procedures in both metric and standard units for hundreds of problems associated with the generation of power by steam, gas, coal, and oil. This handbook is similar in format to another Hicks's work, *Standard Handbook of Engineering Calculations.*

294.  **Switchgear and Control Handbook.** 2nd ed. Edited by Robert W. Smeaton. New York: McGraw-Hill, 1987. 1,056p. illus. index. ISBN 0-07-058449-4. $75.00.

Provides practical information on the design, application, and maintenance of switchgears and controls. Of special interest is the coverage, including interpretations of ANSI, IEEE, NEMA, and UL codes. All aspects of switchgears are covered including programmable controls and solid-state devices.

295.  **Transmission Line Design Manual.** By Holland H. Farr. Denver, Colo.: U.S. Department of the Interior, Water and Power Resources, 1980. 483p. illus. index.

Intended for use by the Bureau of Reclamation in designing power transmission lines, this book has applications for all designers of transmission systems. Coverage includes basic data on poles and lines, conductor sags and tension, insulation, lightning protection and clearance patterns, and guying charts. A bibliography is provided along with appendixes which cover wire data tables, etc.

296.  **Transmission Line Reference Book 345K and Above.** 2nd ed. Palo Alto, Calif.: Electric Power Research Institute, 1982. 180p. illus. $45.00.

Details FHV-UHV transmission systems including electrical characteristics, corona loss, audible noise, field effects of overhead transmission lines, lightning performance, and insulation.

# Radio, Telecommunications, and Television

Books listed here are from the major application of electronics and electrical engineering known collectively as communications. Many engineers are employed designing, manufacturing, servicing, and working in this growing area.

297.  **Antenna Data Reference Manual: Including Dimension Tables.** By Joseph J. Carr. Blue Ridge Summit, Pa.: Tab Books, 1979. 268p. tables. index. LC 79-16886. ISBN 0-8306-9738-1. $13.95.

Tabular presentation of values for antenna lengths which have been previously calculated. Primary coverage is of amateur radio antennas but it also includes citizens band, shortwave, broadcast band, and transmitting antennas.

298.  **Antenna Engineering Handbook.** 2nd ed. Edited by Henry Jasik and Richard Johnson. New York: McGraw-Hill, 1984. various paging. illus. index. LC 83-2867. ISBN 0-07-032291-0. $95.00.

Like the first edition, this handbook is organized into four major parts: introduction and fundamentals, types and design methods, applications, and topics associated with antennas. Many new topics have been added to this classic, including small antennas, microstrip, phased arrays, seeker antennas, and electronic counter methods (ECM). Most chapters include extensive references and bibliographies. Richard

Johnson, editor of the second edition, is a former president of the IEEE Antenna and Propagation Society.

**299. Audio Handbook.** By Gordon J. King. London: Newnes-Butterworths, 1975. 286p. illus. index. ISBN 0-408-00150-x. $26.95.

Covers audio fundamentals, preamplifiers and control circuits, power amplifiers and supplies, loudspeakers and headphones, disc recording and reproduction, microphones and mixers, tape recording, and AM-FM radio. This book is richly illustrated and lists the characteristics of the equipment covered.

**300. Compendium of Communication and Broadcast Satellites 1950-80.** Edited by Martin P. Brown. Piscataway, N.J.: IEEE Press, 1981. 375p. illus. no index. LC 81-81858. ISBN 0-87942-153-3. $40.50.

This text has many illustrations including photographs of each satellite with a person to give a size comparison. The following information on each satellite is given in chart form: the communication payload, frequency plan showing the communications and broadcast bands utilized, listing of all major transmission parameters, physical characteristics, and general information such as launch vehicle, prime contractor, etc. The date of introduction and the designed lifetime are also provided for each satellite.

**301. Handbook of Antenna Design.** Edited by A. W. Rudge, et al. London: Peter Peregrinus, 1983. 2v. illus. index. LC 83-114959. ISBN 0-906048-82-6, 0-906048-87-7. $198.00. (IEE Electromagnetic Waves Series, volumes 15 and 16).

Designed for practicing engineers and students of antenna theory, this illustrated book by 28 contributing authors provides design data and characteristics of radomes.

**302. Modern Radar: Theory, Operation & Maintenance.** 2nd ed. By Edward L. Safford, Jr. Blue Ridge Summit, Pa.: Tab Books, 1981. 564p. illus. index. LC 80-20678. ISBN 0-83-06-9918-x. $21.95.

A basic handbook of radar and radar systems, this text covers the four basic radar systems, applications of radar and related functions including radar-frequency components, tracking and site considerations, receiving systems, space radar, etc.

**303. Practical Aerial Handbook.** 2nd ed. By Gordon J. King. London: Newnes-Butterworths, 1970. 232p. illus. index. LC 78-886715. ISBN 0-408-00001-5.

While essential aerial theory is included, the primary intention of this book is to serve as a useful companion to handbooks on radio, television, and audio engineering. Practical aerial systems are discussed as well as choice, erection, shared aerials, and interference of each system. Appendixes cover aerial dimensions, aerials for color television in the United States and Europe, aerials for stereophonic radio, and lightning protection.

**304. Radio Handbook.** 22nd ed. By William I. Orr. Indianapolis, Ind.: Howard W. Sams, 1982. 1,135p. illus. index. LC 78-64872. ISBN 0-672-21874-7. $39.95.

This classic handbook is popular with both radio amateurs and telecommunications professionals. Coverage includes semiconductor devices, vacuum-tube amplifiers, radio-frequency power amplifiers, single-sideband transmission and reception, frequency synthesis, frequency modulation, radio interference, equipment design, and antenna matching.

305. **Reference Data for Engineers: Radio, Electronics, Computer, and Communications.** 7th ed. Edited by Edward C. Jordan. Indianapolis, Ind.: Howard W. Sams, 1985. various paging. illus. index. LC 43-14665. ISBN 0-672-21563-2. $69.95.

This handbook, a standard for years, is liberally illustrated. It covers filters, attenuators, bridges, rectifiers, transmission lines, and properties of materials used in radio engineering. The book is designed to be used by hobbyists, students, and engineers. Much of the information and data given is of a practical nature. References are found at the end of each chapter.

306. **Reference Manual for Telecommunications.** By Roger L. Freeman. New York: John Wiley, 1984. 1,500p. illus. index. LC 84-13207. ISBN 0-471-86753-5. $75.00.

Gathers into a single, oversized volume: tables, performance curves, nomograms, standards, formulas, constants, interface information, and conversion data. Details the basic discipline of telecommunications including transmission and switching, the related areas of propagation, electromagnetic interference, radio frequency, reliability, etc. Specific topics include telephone traffic, networks and routing, metallic pair systems, coaxial cable, fiber optics transmission, frequency division, multiplex, and pulse codes.

307. **Telecommunication Transmission Handbook.** 2nd ed. By Roger L. Freeman. New York: John Wiley, 1981. 706p. illus. index. LC 81-7499. ISBN 0-471-08029-2. $57.50.

A compilation of information for telecommunication systems engineers which consists of tables, graphs, figures, nomograms, formulas, and statistics. Coverage is broken down into 26 categories each of which has a separate table of contents, references, and, in most cases, a bibliography. Much of the information is from standards or standardizing documents and the source of every entry is listed. This is a truly monumental work. Sections include "Fiber Optics Transmission," "Noise and Modulation, Radio Transmission, Frequency Division Multiples," and "Facsimile Transmission."

308. **Television Engineering Handbook.** Edited by K. Blair Benson. New York: McGraw-Hill, 1986. 1,478p. illus. index. ISBN 0-07-004779-0. $89.50.

This illustrated and indexed handbook provides engineering data for the design, development, maintenance, and operation of television equipment and systems. Coverage includes signal generation, processing, transmission, and reproduction as well as descriptions of the components used in broadcasting, cable, and satellite distribution. Other areas covered are the signal storage disciplines of videotape, disk, and film.

309. **World Radio TV Handbook.** Vol. 1- . New York: Billboard A.C., 1946- . annual. $19.50/yr.

Gives detailed country-by-country information on radio and television stations, including name of the station, address, lists of transmitting stations, frequencies, power and call signs, and program information including language of broadcast country. Of special interest is the list of English broadcasts worldwide with the time, station, and frequency.

# 14   Reference Texts

Reference texts differ from handbooks in the depth of coverage and in the way they are used. Reference texts cover a considerable amount of detail, which means that they are not designed for regular consultation or quick lookup as handbooks are. Rather they are consulted when a handbook does not adequately cover the subject at hand or when more detailed or specialized information is needed.

Reference texts often cover subjects in such detail that they may be used as the text for a course. This does not detract from the usefulness of the book as a reference but in fact often confirms its usefulness.

Because the fields of electrical, electronics, and computer engineering encompass a broad area and represent a large number of books, this section has been divided into smaller sections with somewhat different categories then were used to divide the section on handbooks. "Professional Examination" and "Testing and Troubleshooting" have been added while "Fabrication and Manufacturing" has been dropped. "Basic," "Design," "Power and Power Distribution," and "Radio, Telecommunications, and Television" remain.

## Professional Examination

The books in this category are designed to be consulted and/or studied by engineers prior to taking the Professional Engineering (PE) Examination. Passing this exam is an important professional credential for engineers who wish to act as consultants or set up a private practice.

310. **Electrical Engineering Review Manual.** 4th ed. By Raymond B. Yarbrough. San Carlos, Calif.: Professional Publications, 1983. 416p. illus. index. ISBN 0-932276-36-9. $35.45.

Subtitled "A Complete Review Course for the P.E. Examination for Electrical Engineers," sections cover mathematics, linear circuit analysis, measurements of signal waveforms, time and frequency response, power systems, rotating machines, linear amplifiers, nonlinear waveshaping and switching, digital logic, control systems, illumination, and engineering economic analysis. A copy of a sample examination is also provided. Often used as the text for the review sessions aimed at preparing/updating engineers for the examination, this book is also a good source for examples of worked problems and calculations.

311. **A Programmed Review for Electrical Engineering.** 2nd ed. By James H. Bentley and Karen M. Hess. New York: Van Nostrand Reinhold, 1982. 225p. illus. index. ISBN 0-442-21390-5. $14.95.

This updated edition analysizes and solves the types of problems found in the most recent PEs. Problems are used to illustrate each concept. Especially helpful are the graphic aids and alternative solutions. Another useful part is the coverage of digital logic and engineering economics.

# Basic Texts

312. **Basics of Electricity and Electronics.** 2nd ed. By Matthew Mandl. Englewood Cliffs, N.J.: Prentice-Hall, 1975. 448p. illus. index. ISBN 0-13-060228-0. $27.95.

Formerly *Fundamentals of Electric and Electronic Circuits*, this new edition covers the basic principles of electricity and electronics. The book is aimed at users who need or would like a comprehensive discussion of the relationship between electricity and electronics. Some of the areas covered include simple series and parallel circuits, complex series and parallel circuits, principles of magnetism, DC measurements, resonance, transformers, and capacitance. The author has written a number of well-known books on electronics.

313. **Digital Electronics: Logic and Systems.** 2nd ed. By John K. Kershaw. Belmont, Calif.: Breton Publishers, 1983. 546p. illus. index. LC 82-17891. ISBN 0-534-01471-2. $25.00.

Because this book was designed to be used as a community college textbook, each chapter is headed by a set of objectives. It includes not only the TTL and CMOS families, but also memory circuits, RAMs, ROMs, microprocessors, and the applications of these devices. Added to the second edition is a chapter on multiplexers, demultiplexers, and programmable logic arrays. Glossaries, which are included to support the text, are arranged in "logical" order rather than alphabetical order. This logical order has a somewhat negative effect on their usefulness because while it does place like concepts and terms together, it is somewhat difficult to find the definition of a term if it is out of alphabetical order and placed with like terms.

314. **Fundamentals Handbook of Electrical and Computer Engineering.** Edited by Sheldon S. L. Chang. New York: Wiley-Interscience Publication, 1982-1983. 3v. index. LC 82-4872. ISBN 0-471-89690-X. $189.95/set.

This three-volume set consists of *Circuits, Fields and Electronics, Communications, Control, Devices and Systems,* and *Computer Hardware, Software and Applications.* Each volume may be purchased and used separately as each has its own table of contents, list of contributors, and index. The stated aim of this set is to "constitute a coherent, concise treatise of the core areas of electrical and computer engineering with emphasis on system, device and circuit design."

**315. Guide to CMOS Basics, Circuits & Experiments.** By Howard M. Berlin. Indianapolis, Ind.: Howard W. Sams, 1979. 221p. illus. index. LC 79-67128. ISBN 0-6722-1654x. $9.95.

A heavily illustrated book which is popular with students. This book covers the nature of CMOS, basic CMOS devices, CMOS characteristics, and design rules. Of particular interest are the description of 22 experiments using CMOS devices. The appendixes, which cover TTL to CMOS conversion and breadboarding aids, alone make this book valuable.

**316. Handbook for Transistors.** By John D. Lenk. Englewood Cliffs, N.J.: Prentice-Hall, 1976. 296p. illus. index. LC 75-2294. ISBN 0-13-3822672. $7.95.

One of many written by Lenk, this book consists of three areas: basic theory, analysis of components and circuits, and proven design practices. It is intended to be used as a "cookbook." Many users will appreciate one of the strongest parts of the book—the explanation of how to use the information found on commercial data sheets, including how to test specific device characteristics.

**317. Handbook of Digital Electronics.** By John D. Lenk. Englewood Cliffs, N.J.: Prentice-Hall, 1981. 384p. illus. index. LC 80-19647. ISBN 0-13-377184-9. $24.95.

The intent of this book is to provide users with both a "crash course" in digital electronics and the basic approaches to digital testing and troubleshooting. Coverage includes a review of the numeric and alphanumeric codes used; digital logic including gates, decoders, and multiplexers; and cross sections of digital circuits. Special emphasis is placed on ICs and the elements of a typical digital computer. The last section covers test procedures and troubleshooting. Like other books by Lenk, this is a clearly written book filled with useful information and technical details.

**318. Handbook of Radar Measurement.** 2nd ed. By David K. Barton and Harold R. Ward. Dedham, Mass.: Artech House, 1984. 415p. illus. index. ISBN 0-89006-155-6. $25.95.

An updated version of Barton and Ward's *Handbook*, this book features the theory, equations, tables, and graphic information needed to estimate radar range accuracy, Doppler frequency, and angle measurements. Contents include angular measurement in noise, range, and Doppler measurement; multiple-signal problems; target-induced errors; and error analysis. The appendixes include antenna patterns and illumination functions, waveform analogies, data filters and smoothing, atmospheric propagation errors, and a bibliography.

**319. IC Array Cookbook.** By Walter G. Jung. Rochelle Park, N.J.: Hayden Book Co., 1980. 208p. illus. index. LC 79-2792-3. ISBN 0-8104-0762-0. $9.75.

Like the *IC Op Amp Cookbook, IC Timer Cookbook,* and *IC Converter Cookbook,* this is an extremely practical book, which is very popular with students. It is divided into three parts. The first part includes an introduction covering the basic concepts and the range of IC arrays available. The second part covers circuit applications like AC and DC amplifiers, gain controls, and regulators. The third part details selected data sheets, manufacturers source information, device cross-references, and a listing of distributors. All circuits are presented with specific component values, description of operation, range, limits, and suggested modifications. References are listed at the end of each chapter.

320. **Introduction to Microprocessors: Experiments in Digital Technology.** By Noel T. Smith. Rochelle Park, N.J.: Hayden Book Co., 1981. 176p. illus. index. LC 81-6568. ISBN 0-8104-0867-8. $12.95.

Designed to teach the use of integrated circuits from the simplest gates to multibit microprocessors, this book provides an explanation of the elemental parts that give ICs their capabilities and characteristics. The appendixes cover Melton's special breadboard, IC type number identification, in-out diagrams, and other schematics and charts designed to provide users with an understanding of ICs and the capability to use ICs in electronic design.

321. **Intuitive IC Electronics.** By Thomas M. Frederiksen. New York: McGraw-Hill, 1981. 208p. illus. index. LC 80-27910. ISBN 0-07-021923-0. $19.95. (National Semiconductor Technology Series).

Very clear introduction to the operating principles of IC electronics. National Semiconductor, one of the major manufacturers of ICs in the United States, was responsible for much of the early research and development.

322. **Microprocessors.** 3rd ed. By Rodnay Zaks. Berkeley, Calif.: Sybex, 1980. 413p. illus. index. LC 80-51035. ISBN 0-89588-042-3.

Covers fundamental concepts like internal operations, systems components, comparative microprocessor evaluations, system interconnects, microprocessor applications, interfacing techniques, and microprocessor programming in a clear concise manner. The appendixes include electronic symbols and the instruction sets for the Motorola 6800, Intel 8080, and the S-100 bus. The inclusion of the latter adds to the usefulness of the book as users are often looking for these instruction sets for devices found in existing equipment. A bibliography follows the index.

323. **Physics of Semiconductor Devices.** 2nd ed. By S. M. Sze. New York: Wiley-Interscience Publication, 1981. 868p. index. LC 82-213. ISBN 0-471-05661-8. $50.95.

One of the major text and reference books available on the subject of semiconductors, this second edition has been extensively updated—over 80 percent of the material is new or has been updated. Of the 1,000 references, over 70 percent were published since 1960. Like the world of semiconductors, this book is divided into bipolar, unipolar, microwave, and photonic devices. Each section includes a historical perspective and gives up-to-date properties of the devices covered. There are extensive references at the end of each section. The appendixes include physical and lattice constants and the properties of semiconductor materials like gallium, silicon, and gallium-arsinide. This book should be in any collection that includes solid-state physics or electronics.

**324. TTL Cookbook: Complete Guide to Understanding and Using Transistor Logic.**
By Donald E. Lancaster. Indianapolis, Ind.: Howard W. Sams, 1974. 335p. illus. index.
LC 73-90295. ISBN 0672-210355. $12.95.

Like the author's *RTL Cookbook,* this book covers in simple language the basics of
TTL (transistor to transistor logic), including who makes it, where to get data, how to
interconnect, and how to power the device. Chapter 3 covers logic, while chapter 5
details gates and timers. Chapter 6 describes dividing by N counters, chapter 7 shift
registers, noise generators, and rate multiplexers. Applications information is provided
in chapter 8. The appendix gives a list of manufacturers with addresses for the devices
discussed.

# Design

Books in this group are designed to answer the kinds of questions engineers would
have as they seek to create systems, devices, and processes that solve technical
problems. Often the kind of information needed concerns components and peripherals.

**325. Active Filter Cookbook.** By Don Lancaster. Indianapolis, Ind.: Howard S.
Sams, 1975. 240p. illus. index. LC 74-33839. ISBN 0-672-21168-8. $14.95.

A richly illustrated, popular book which deals with circuits that can be easily
designed, built, and tuned. Covers the so-called "ripoff" circuits. The writing is clear
and descriptive. What sets this book apart from others is the discussion of why a circuit
won't work in a particular application. Another useful feature is the catalog of commercially available circuits. This is a good stand-alone discussion of active filters.

**326. Connectors and Interconnections Handbook.** Edited by Gerald L. Ginsberg.
Camden, N.J.: Electronic Connector Study Group, 1977. various paging. illus. index.
LC 77-2000.

Covers connectors, wiring and cabling, terminations, materials, and standards and
specifications. Separate title word and subject indexes are included along with a glossary
of connection, printed wiring and wire, and cable terminology. The book is generously
illustrated with drawings of techniques and property charts. Connectors are an
important area in electronic design and this book, one of the few available on the topic,
gives good detail on their current technology.

**327. Counting and Counters.** By R. M. Oberman. New York: John Wiley, 1981.
192p. illus. index. LC 80-26966. ISBN 0-470-27118-3. $34.95.

The author considers "any circuit running through a cycle of states to be a
counter." While designed as a graduate text, this book also serves as a reference and
explanation of commercially available counter circuits. Among the counters covered are
binary and constant-ratio counters, accumulative and shift register counters, and
pseudo-random generators. A short reference list accompanies each chapter.

**328. CRT Controller Handbook.** By Gerry Kane. New York: Osborne/McGraw-Hill,
1980. 224p. illus. index. LC 82-221871. ISBN 0-931988-45-4. $9.95.

Describes all of the CRT (cathode ray tube) controller devices available up to 1980.
Covers pins, signals, programmable registers, microprocessor interfaces, screen
memories, transparent memories addressing, and character generator interfaces. Illustrated, it includes a clear explanation of the principles of CRT operation. CRT

technology is an important component of many electronic devices and instruments, not the least of which are computers and oscilloscopes.

329. **Design and Drafting of Printed Circuits.** 2nd ed. By Darryl Lindsey. Westlake Village, Calif.: Bishop Graphics, Inc., 1983. 440p. illus. index. LC 81-69173. ISBN 0-07-037844-4. $45.95.

Written by the founder of a printed circuit design school, almost half of this book has been added or revised since the first edition was published in 1979. Illustrated with schematic drawings, the intention is to explain the how, why, and use of printed circuits. Designed as both a teaching and reference tool, the book deals with concepts like multilayer boards, flexible circuits, CAD (Computer aided design), and automatic assembly. Chapter 11 has a sequence of photographs illustrating the step-by-step process of artmaster tapeup.

330. **Display Devices.** Edited by J. I. Pankove. New York: Springer-Verlag, 1980. 300p. illus. index. LC 79-27608. ISBN 0-387-09868-2. $53.00. (Topics in Applied Physics, vol. 40).

Intended for both graduate students and practicing engineers, the aim of this book is to present basic information about the most important types of display devices and at the same time give limited information about the more esoteric devices. The display devices in the text are grouped by whether they emit or modulate light. It is intended for either practicing engineers or graduate students who need information on a rather specialized area. References follow each chapter.

331. **Display Electronics.** By Ken Tractor. Blue Ridge Summit, Pa.: Tab Books, 1977. 252p. illus. index. LC 77-79348. ISBN 0-8306-6861-1. $6.95.

This is a good basic explanation and how-to manual about display devices, which are a mainstay of electronics. These devices allow users to make observations and interpret data. Devices covered include photon emission, transmission, detection, light emitting diodes, arrays and displays, electroluminescence, and injection laser diodes and applications.

332. **Functional Circuits and Oscillators.** By Herbert J. Reich. Princeton, N.J.: Van Nostrand Reinhold, 1961. 466p. illus. index. LC 62-19117. (Van Nostrand Series in Electronics and Communications).

Intended to serve as both a text and reference to feedback oscillators and negative resistance devices. Coverage includes decade ring and glow diode, counting circuits, sawtooth voltage generators, Eccles-Jordan circuits, and Colpitts oscillators.

333. **Handbook of Active Filters.** Edited by David E. Johnson, et al. Englewood Cliffs, N.J.: Prentice-Hall, 1980. 244p. illus. index. LC 77-10373. ISBN 0-13-372409-3. $28.95.

Details the basic filters such as low pass, high pass, band pass and band reject, and Butterworth and Cheyshev, along with infinite gain and IC operational amplifiers. A good general reference source on an important area.

334. **Handbook of Electric Machines.** Edited by Syed A. Nasar. New York: McGraw-Hill, 1987. 608p. illus. index. ISBN 0-07-045888-x. $59.95.

Provides comprehensive coverage of all the major facets of electric machines. Detailed coverage is given of large synchronous, induction, and DC machines. Smaller machines such as single-phase, electronic control, and linear motors are also covered. Each machine type includes physical description, equivalent circuits, performance calculations, design equations, winding details, and standards for ratings and testing.

335. **Handbook of Electric Motors: Use and Repair.** By John E. Traister. Englewood Cliffs, N.J.: Prentice-Hall, 1984. 272p. illus. index. LC 83-24765. ISBN 0-13-377383-3. $24.95.

Written to meet the needs of the electric shop and design engineers, this book is not intended to be a "crash course" in electric motors. Coverage includes split-phase motors, capacitor motors, repulsion type, and AC and DC motor control, DC armature windings, installation, and selection. The book is especially strong in the area of motor selection. Examples of motors used in a number of different applications are given. While early electrical engineers were given extensive training about motors, engineering students are now given limited exposure at best. Therefore, this book is important in that it does an excellent job of covering current motors.

336. **Handbook of Electronics Industry Cost Estimating Data.** By Theodore Taylor. New York: John Wiley, 1985. 410p. illus. index. LC 85-6294. ISBN 0-417-82264-7. $59.50.

Originally published in 1964 by Fred C. Martmeyer, this book has been revised, updated, and expanded. Coverage includes machining and sheet metal fabrication; wiring, circuit board, and integrated component assembly; electrical testing; and packaging. The book consists of time standards, manufacturing methods, and "rules of thumb" which are used for cost-estimating electronic equipment and systems. Data in this book were based on individual interviews, onsite inspections, and company experience.

337. **Handbook of Machine Soldering: A Guide for the Soldering of Electronic Printed Wiring Assemblies.** By Ralph W. Woodgate. New York: John Wiley, 1983. 224p. illus. index. LC 82-17540. ISBN 0-471-87540-6. $27.50.

The focus of this book is the creation of zero-defect soldering. To achieve this end, the book covers choosing, installing, and operating soldering machines. It also explains the necessary theory and gives detailed descriptions of available machines and systems so that the designer can achieve defect-free wave soldering. Soldering is a very important aspect of electronics and one that is not often covered in much detail in textbooks.

338. **Handbook of Operational Amplifier Circuit Design.** By David F. Stout and Milton Kaufman. New York: McGraw-Hill, 1976. various paging. illus. index. LC 76-3491. ISBN 0-07-061797-x.

Provides design information on 68 op-amp circuits. Equations are presented in tabular form with references at the end of each section. Each circuit is accompanied by a parameter list and a set of design equations. The intent is to give step-by-step instructions for designing or analyzing a circuit using an op-amp. The appendixes cover decibel calculations, PC circuit characteristics, and circuit fabrication. This book's strong point is its lists of the design parameters and equations with each circuit.

**339. Handbook of Semiconductor and Bubble Memories.** By Walter A. Triebel and Alfred E. Chu. Englewood Cliffs, N.J.: Prentice-Hall, 1982. 401p. index. LC 81-2468. ISBN 0-13-381251-0. $26.95.

Designed to serve as a reference for technicians and engineers, the first part of this book is an introduction to electronic memories, giving the basics and characteristics of each type of memory. The memories covered include read-only and random-access, read/write to charge-coupled devices, and magnetic bubbles. The bibliography is extensive, with the citations coming primarily from company data sheets and application notes.

**340. Handbook of Simplified Solid State Circuit Design.** 2nd ed. By John D. Lenk. Englewood Cliffs, N.J.: Prentice-Hall, 1978. 429p. index. LC 77-23555. ISBN 0-13-381715-6. $24.95.

Like the author's other books, this is a step-by-step approach to solid-state circuit design. The text assumes no previous design experience. Covers basic transistor bias, audio amplifiers, transformers, operation amplifiers, radio frequency circuits, wave forming and waveshaping circuits, and power supply.

**341. Hybrid Microcircuit Reliability Data: A Practical Sourcebook for Designers, Fabricators & Users.** Edited by IIT Research Institute. New York: Pergamon Press, 1976. 204p. tables. LC 75-29637. (Classic Handbook Reissue Series).

Consists of highly summarized but detailed test and operation data for use in part selection, failure rate predictions, screening decisions, environmental test specifications, and preparation and failure models.

**342. Hybrid Microcircuits.** By T. D. Towers. New York: Crane Russak, 1977. 246p. illus. index. LC 76-53092. ISBN 0-7273-0801-7.

Designed as a users' guide to hybrid integrated circuits, this book covers special user features, manufacturers and packages, passive hybrid and high power hybrids, precision and RF hybrids, and optoelectronic and custom-built circuits. References are found in the appendixes along with lists of manufacturers and suppliers of parts and equipment.

**343. Logic Designers Guidebook.** By E. A. Parr. New York: McGraw-Hill, 1984. 480p. illus. index. LC 84-7864. ISBN 0-07-048492-9. $42.50.

Aims to provide an accessible source of data on devices in the TTL and CMOS families. Details combinational logic, storage, timers, counters and shift registers, binary arithmetic, communications circuits, and hardware information. Essential references are given for each circuit. Unlike most McGraw-Hill books, however, no references to other sources of information are provided.

**344. Manual for MOS Users.** By John Lenk. Reston, Va.: Reston Publishing, 1975. 340p. illus. index. LC 74-23929. ISBN 0-87909-478-8.

Designed for users of metal oxide semiconductor (MOS) devices rather than designers, this book uses existing commercial MOS devices. It covers working with discrete MOS devices, and differential and RF amplifiers. Discusses how to use data sheets and other manufacturers data and information.

345. **Manual for Operational Amplifier Users.** By John D. Lenk. Reston, Va.: Reston Publishing, 1976. 292p. index. LC 75-33964. ISBN 0-87909-477-x.

Intended as a practical reference for users and designers of op-amps. As a result, the handbook focuses on commercially existing op-amps to solve design and application problems. It has three stated purposes: to acquaint the reader with op-amps, to provide basic op-amp characteristics, and to encourage use of op-amps. Like other books by Lenk, this is very clearly written and enormously popular with students.

346. **Manual of Active Filter Design.** 2nd ed. By John L. Hilburn and David E. Johnson. New York: McGraw-Hill, 1983. 256p. illus. index. LC 82-4649. ISBN 0-07-028769-4. $39.95.

Distinguished from several other handbooks on active filters by the extensive use of design graphs like K parameter versus frequency or second-order biquad low-pass. The five types of filters are each discussed in a separate chapter, with design procedure and graphs found at the end of that chapter.

347. **Master Handbook of Microprocessor Chips.** By Charles K. Adams. Blue Ridge Summit, Pa.: Tab Books, 1981. 378p. illus. index. LC 80-28678. ISBN 0-8306-9633-4. $18.95.

Made up primarily of copies of data sheets from Intel, Mostek, and Texas Instruments, this book is organized by families of chips. For example: 4004 and 4040, 8008 and 8080, MC6800, Z80/Z80A, 16-Bit microprocessor, and EPROM. The operation of the chips is covered in detail along with the relevant instruction sets.

348. **Microelectronics: Standard Manual and Guide.** By John Douglas-Young. Englewood Cliffs, N.J.: Prentice-Hall, 1983. 277p. illus. index. LC 83-9450. ISBN 0-13-581108-2. $21.95.

Covers both analog and digital applications of the two major microelectronic technologies—bipolar and unipolar. The detailed discussions are aimed at technicians with a lot of analog and discrete theory experience, but who are not yet familiar with digital theory. The appendixes include a glossary of terms; abbreviations; a good selection of electronic formulas and mathematics tables; graphic symbols for semiconductor devices; and an excellent explanation of the binary number system. A good basic book by the author of several well-known electronic reference books.

349. **Microprocessor Application Handbook.** By David F. Stout. New York: McGraw-Hill, 1982. 448p. illus. index. LC 81-11787. ISBN 007-0617988. $39.95.

Presents a variety of microprocessor applications using the following types: 1802, 2650, 6800, and 8080, as well as 2920, 3870, 3872, 6801, 6802, and 8088. LSI instruments are discussed including analog-to-digital, digital-to-analog converters, communications devices, timers, clocks, and video interfaces. The IEEE 488 bus is also covered. References are found at the end of most chapters. Like most McGraw-Hill handbooks, each chapter was written by an expert. This book is very popular with users.

350. **Practical Guide to Digital Integrated Circuits.** 2nd ed. By Alfred W. Barber. Englewood Cliffs, N.J.: Prentice-Hall, 1984. 272p. illus. index. LC 83-21208. ISBN 0-13-690751-2. $21.95.

A helpful, realistic guide to digital ICs, this book covers the relationship between ICs and transistors with explanations of how different circuits have been used in logical ICs and flip-flops. Other areas covered are how to combine gates and flip-flops, use of breadboards, design of complex systems using the basic elements of electronics, ways to analyze large scale integration, and tips on selecting and using ICs. Of special interest is the discussion on ways to test ICs.

351. **Printed Circuit Boards for Microelectronics.** 2nd ed. By J. A. Scarlett. New York: State Mutual Book, 1980. 330p. illus. index. LC 77-105347. ISBN 0-901150-08-8. $145.75.

Concerned with the design of the printed circuit boards (PCB) that are used for mounting integrated circuits, what sets this book apart is the section "Examples of Artwork." Here photographs and drawings clearly show possible variations in the artwork. Artwork is discussed in many books, but few books have quality photographs which illustrate the specific types and steps used to create the artwork on a PCB.

352. **Quick Reference Manual for Silicon Integrated Circuit Technology.** Edited by W. E. Beadle, J. C. Tsai, and R. D. Plummer. New York: John Wiley, 1985. 736p. illus. index. LC 84-25667. ISBN 0-471-81588-8. $65.00.

A compendium of graphs, charts, recipes, nomographs and other reference data which have been collected by engineers and scientists of AT&T Bell Laboratories. Topics covered include properties of silicon, mathematical expressions, diffusion ion implantation, conductivity of diffused layers, properties of p-n and metal semiconductor junctions, and MOS. Because the charts are intended for quick lookup, the explanations are terse. The information included is designed to augment designers basic knowledge rather than to educate them. This collection should be very useful for engineers in the field and graduate students specializing in design problems.

353. **RC Active Filter Design Handbook.** By F. W. Stephenson. New York: John Wiley, 1985. 480p. illus. index. LC 84-23707. ISBN 0-471-86151-0. $44.95.

Written especially for nonspecialists, this handbook's practical approach will be much appreciated by students. It covers design rules, gives hints on construction and measurement of filters, and provides approximation data so that a wide range of filter characteristics may be designed. Also included is the code for a computer program for use on home computers to aid in designing active filters.

354. **Talking Chips.** By Nelson Morgan. New York: McGraw-Hill, 1984. 224p. illus. index. LC 83-14914. ISBN 0-07-043107-8. $25.95.

This conversational book has some very useful information. Intended as a bridge between graduate level texts and product literature, it includes an overview of speech synthesis and hardware, an abbreviated discussion of phonetics, and a glossary of terms, as well as a description of how chips make intelligible sounds.

355. **Transistors: Fundamentals for the Integrated Circuit Engineer.** By R. M. Warner and B. L. Grung. New York: Wiley-Interscience Publication, 1983. 875p. illus. index. LC 83-10392. ISBN 0-471-09208-8. $64.95.

"Highly recommended for libraries serving graduate solid-state physicists and integrated circuit engineers, particularly for its authoritative contents and extensive references to pertinent technical and periodical literature." (*Choice*, April 1984, p. 1163).

356. **A User's Handbook of Semiconductor Memories.** By Eugene R. Hnatek. New York: Wiley-Interscience Publication, 1977. 652p. index. LC 77-362. ISBN 0-471-40112-9. $61.95.

A comprehensive overview of memory devices including their advantages and characteristics. Coverage includes bipolar, MOS, shift registers, programmable logic arrays, and charge coupled devices.

357. **Video and Electronic Displays: A User's Guide.** By Sol Sherr. New York: Wiley-Interscience Publication, 1982. 352p. illus. index. LC 81-21915. ISBN 0-471-09037-9. $34.95.

Designed for users who do not have a strong technical background, but do need information about electronic displays. This book is divided into two parts, the first of which deals with the equipment, while the second covers equipment applications. Areas detailed include electronic display systems, human factors, television or raster scan displays, vector and raster full graphics, and flat-panel displays, as well as the applications of alphanumeric, numeric, and computer graphics systems. Also has a short section on how to do performance evaluations on equipment and systems. A short bibliography is included.

## Power and Power Distribution

358. **Advanced Power Cable Technology.** By Toshikatsu Tanaka and Allan Greenwood. Cleveland, Ohio: CRC Press, 1983. 2v. illus. index. LC 82-14597. ISBN 0-8493-5165-0, 0-8493-5166-9. $68.00 and $78.00, respectively.

Volume 1 covers the basic concepts of power cables including the materials used in cable technology, while volume 2 details current cable technology, future cables, and cable improvement. References are provided at the end of volume 2. Little emphasis is given to commercially manufactured cables, because the intent is to give an overview of the basic types of cables and their properties and application. This is one of the few books devoted to the important area of cable technology.

359. **Basic Electrical Power Engineering.** By Ollie I. Elgerd. Reading, Mass.: Addison-Wesley, 1977. 494p. illus. index. LC 76-1751. ISBN 0-201-01717-2.

Covers the entire area of electric power from generation, transformation, and transmission to the final conversion processes in electric motors. The text makes extensive use of examples and analogies. Among the chapters of interest are "Synchronous Electric Energy Converters," "Power Transformers," "Electric Power Networks," "Phasor Analysis," and "Spectral Analysis." Answers to selected problems are included at the end of the text.

360. **Electric Power Transmission Systems.** 2nd ed. By J. Robert Eaton and Edwin Cohen. Englewood Cliffs, N.J.: Prentice-Hall, 1983. 432p. illus. index. LC 82-21514. ISBN 0-13-247304-6. $26.95.

Designed to cover the technical aspects of the electric systems which transmit power from the generators to the loads on the system, this book should be of use to technicians who build, operate and maintain electric power equipment, to students who need more advanced study, and people who are associated with the electric power industry. Chapters include "Percent and Per Unit Quantities," "Assemblies of Power System Components," "Power Limits," "Faults on Power Systems," "Relays and Relay Systems," "Electrical Insulation," and "Power Distribution."

361. **Electric Utility Systems and Practices.** 4th ed. Edited by Homer M. Rustebakke. New York: Wiley-Interscience Publication, 1983. 336p. illus. index. LC 83-3640. ISBN 0-471-04890-9. $42.50.

Provides an excellent description of how power system components are designed and operated. Chapters include information on power generation, power from combustion turbines, power from water, transmission, transformers, switchgear, and protective relaying. References are given at the end of most chapters. This book, prepared by the Electric Utility Systems Engineering Department of General Electric for use by their engineers and sales personnel, is one of the few books published recently which deals with modern utility systems.

362. **Handbook of Power Generation: Transformers and Generators.** By John E. Traister. Englewood Cliffs, N.J.: Prentice-Hall, 1982. 272p. illus. index. LC 82-20480. ISBN 0-13-380816-5. $21.95.

The text details principles and characteristics of DC and AC generators, engine and gas turbine driven generators, conversion equipment, primary and secondary distribution systems, transformer control, installation, construction, characteristics and mounting, pole line construction, and underground wiring. Emphasis is on practical, on-the-job applications. A glossary of terms is included.

363. **Handbook of Practical Electrical Design.** Edited by J. F. McPartland. New York: McGraw-Hill, 1984. various paging. index. LC 82-20798. ISBN 0-07-045695. $31.95.

This handbook is directed at electrical personnel and mechanical engineers involved with the design of electrical applications. Coverage includes practical procedures for design of electrical circuits and systems to supply lighting motors, air conditioning, signals, and controls in industrial, commercial, and residential applications.

364. **Handbook of Simplified Electrical Wiring Design.** By John D. Lenk. Englewood Cliffs, N.J.: Prentice-Hall, 1975. 416p. illus. index. LC 74-9998. ISBN 0-13-381723-7. $24.95.

This practical handbook is aimed at helping electricians and contractors solve specific problems by analyzing the basic capabilities and limitations of different wiring systems. Specifically includes the basics of power distribution, raceways and conductors, grounding systems, transformers, heating, and electric motor control.

365. **Power Generation Operation and Control.** By Allen J. Wood and Bruce F. Wollenberg. New York: John Wiley, 1984. 444p. illus. index. LC 83-1172. ISBN 0-471-09182-0. $36.95.

The purpose of this book is to discuss and explore both the engineering and economic factors involved in planning, operating, and controlling the power generation and transmission systems used by electric utilities. Chapters include "Characteristics of Power Generation Units," "Economic Dispatch of Thermal Units," "Transmission Losses," "Unit Commitment," "Generation with Limited Energy Supply," "Hydrothermal Coordination," "Control of Generation," "Interchange Evaluation and Power Pools," and "Power System Security." Each chapter is followed by an appendix, problems, and lists of further reading. This book is very popular with students.

366. **Power System Harmonics.** By J. Arrillaga and D. A. Bradley. New York: John Wiley, 1985. 336p. illus. index. LC 84-22097. ISBN 0-471-90640-9. $39.95.

For students as well as researchers, this book covers power system waveform distortion, which is also known as harmonics. The major areas detailed are the causes, effects, analysis, monitoring, penetration, and control of harmonics. A limited discussion of standards and computer models is included. Good coverage of an important area.

367. **Standard Application of Electrical Details.** By Jerome F. Mueller. New York: McGraw-Hill, 1984. 313p. illus. index. LC 83-782. ISBN 0-07-043961-3. $42.50.

The purpose of this book is to supplement existing design texts with a collection of drawings which show electrical detail. The author has chosen those electrical details most commonly used by designers in the following four areas: lighting, power, distribution, and systems and control. Each area includes information on safety, capacity, flexibility, accessibility, and reliability. No coverage is given to outlets.

## Radio, Telecommunications, and Television

368. **AM-FM Broadcasting Equipment, Operations, and Maintenance.** By Harold E. Ennes. Indianapolis, Ind.: Howard W. Sams, 1974. 800p. illus. index. LC 74-75274. ISBN 0-672-21012-6.

This book is concerned with basic information and data needed by the broadcast engineer. Areas covered include semiconductors and logic for broadcast engineers, monaural studios and control rooms, mobile and field facilities, AM and FM transmitters and antenna systems, studio operations, remote pickup operations, and studio maintenance.

369. **Audio Control Handbook: For Radio and Television Broadcasting.** 5th ed. By Robert S. Oringel. New York: Communication Arts Books, 1983. 380p. illus. index. LC 83-6131. ISBN 0-8038-0550-0. $14.95.

Details the operation of the different types of electronic equipment found in the studios and control rooms of radio or television stations. Includes information on control boards, console facilities, microphones, and remote broadcasts. This fifth edition consists of a major revision of the fourth, which was published in 1972.

370. **Care and Feeding of Power Grid Tubes.** By Robert I. Sutherland. Wilton, Conn.: Radio Publications, Inc., 1967. 158p. illus. index. ISBN 0-933616-06-6. $7.95.

This handbook analyzes the operation of power grid tubes and provides design and application data for long tube life and maximum circuit stability.

371. **Complete Home Video Book: A Source of Information Essential to the Video Enthusiast.** By Peter Utz. Englewood Cliffs, N.J.: Prentice-Hall, 1983. 2v. 608p. illus. index. LC 82-10129. ISBN 0-13-161364-2. $29.95/set.

A basic introduction to video requiring no prior electronic or photographic experience, this book covers maintenance of heads and cameras, as well as information about television sets, antennas, cable, and recorders. It also deals with nonelectronics areas like production techniques, including scripting, copying, and lighting. The appendixes list periodicals, books, and tapes of interest along with relevant manufacturers and their addresses.

372. **Designing and Maintaining the CATV and Small TV Studio.** 2nd ed. By Kenneth Knecht. Blue Ridge Summit, Pa.: Tab Books, 1976. 281p. illus. index. LC 72-87450. ISBN 0-8306-6815-2.

Illustrated with numerous pictures of available equipment, such as dimmer systems and three input multiplexers. Also covers studio pulse systems, switching and special effect systems, cameras and lighting equipment, film chain, video recorders, color equipment, RF and video monitors, distribution systems, audio mixing consoles, and program sources.

373. **Handbook of Electronic Systems Design.** By Frank Weller. Reston, Va.: Reston Publishing, 1979. 288p. illus. index. LC 77-10998. ISBN 0-87909-322-6. $57.50.

Designed to fill the gaps between academic circuit theory, system design, and engineering techniques, this book deals with the basic characteristics of telephone and radio systems, radar systems, basic telemetry systems, high fidelity systems, CATV, and production control systems. It is liberally illustrated with diagrams, schematics, and photographs of equipment. Appendixes cover color codes, capacitor codes, characteristics of series, and parallel-resonant circuits.

374. **Handbook of Microwave Techniques and Equipment.** By Harry E. Thomas. Englewood Cliffs, N.J.: Prentice-Hall, 1972. 319p. illus. index. LC 72-2347. ISBN 0-13-380329-3.

The heavy use of tables makes this book popular with students. Of special interest are the sections on impedance matching and Smith charts, component measurement and equipment, microwave antennas. The 27 appendixes, which are primarily tabular, cover microwave terms and definitions, mismatch loss charts, attenuation in coaxial cables, and VSWR calculation nomographs.

375. **Man-Made Radio Noise.** By Edward M. Skomal. New York: Van Nostrand Reinhold, 1978. 350p. illus. index. LC 77-19109. ISBN 0-442-27648-6. $32.50.

Discusses the noise caused by industrial and consumer products and defines the difference between man-made and naturally occurring radio noise. Data are provided to help the user make predictions concerning average power, quasi-peak, and peak noise field intensity requirements.

376. **Microphone Handbook.** By John Eargle. Plainview, N.Y.: Elar Publishing, 1982. 231p. illus. index. LC 81-70852. ISBN 0-914130-02-1. $31.95.

This work is divided into four sections: fundamentals of microphone design, including sensitivity ratings and directional work characteristics; microphones in the physical environment, including proximity and distance effects; microphones in the aesthetic environment, like the studio or speech and music reinforcement; and practical topics such as accessories and interface problems. A bibliography by subject is included.

377. **Practical RF Design Manual.** By Doug DeMaw. Englewood Cliffs, N.J.: Prentice-Hall, 1982. 288p. illus. index. LC 81-14378. ISBN 0-13-693754-3. $27.95.

A practical, nontheoretical treatment of radio frequency (RF) transistors, and ICs. Each circuit is described in detail so that the circuit can be used as a building block for creating composite systems in RF communications. Topics covered include transmitter and receiver fundamentals, frequency control systems, small signal RF amplifiers, large signal amplifiers, and IR amplifiers.

378.   **Satellite Communications.** By Stan Prentiss. Blue Ridge Summit, Pa.: Tab Books, 1983. 288p. illus. index. LC 83-4844. ISBN 0-8306-0632-7. $16.95.

The various disciplines of satellite communications are detailed, including uplink/downlink transceivers and earth stations. Other related areas such as CATV and security and scrambling devices are also included.

379.   **Television Service Manual.** 5th ed. By Robert G. Middleton. Indianapolis, Ind.: Theodore Audel, 1984. 501p. illus. LC 83-22313. ISBN 0-672-23395-9. $15.95.

Treats the entire scope of video transmission, reception, system and circuit theory, and installation and maintenance. Aimed at the audience wanting/needing "a complete study and reference handbook for television." Antenna arrays and transmission lines for color televisions are dealt with in considerable detail. A glossary follows the text.

380.   **Video Electronics Technology.** By Dave Ingram. Blue Ridge Summit, Pa.: Tab Books, 1983. 250p. illus. index. LC 82-5959. ISBN 0-8306-2474-0. $15.95.

Covers black and white television receivers, color television operations and receivers, television antenna, MDS concepts, satellite television receiver systems, home TVRO equipment, video recording, and television troubleshooting. This book provides a straightforward discussion of television systems and related areas including satellite television systems, microwaves, and video playback.

# Testing and Troubleshooting

Troubleshooting is defined as the process of finding electrical or mechanical faults in a system. The concept of troubleshooting assumes that the system was performing properly at one time. Testing and troubleshooting are important areas for technicians, designers, and fabricators because all electronic systems develop "bugs" that need to be either repaired or eliminated. The following books discuss techniques and equipment useful in discovering and correcting electronic bugs.

381.   **Electronic Measurement and Instrumentation.** By Bernard M. Oliver and John M. Cage. New York: McGraw-Hill, 1971. 720p. illus. index. LC 71-124141, ISBN 0-07-047650-0. $62.50. (Intrauniversity Electronics Series, vol. 12).

Details instruments that are neither obvious nor esoteric. Some of the areas covered are sinewave testing, square wave and pulse testing, signal analysis by digital techniques, voltage and current measurements, impedance measurements, and audio frequency signal source measurements. Equipment types covered in detail include oscilloscopes, recorders, audio and video amplifiers, microwave signal analysis receivers, etc. This is a popular book because of its content and the presentation.

382.   **Electronic Meters: Techniques and Troubleshooting.** By Miles Ritter-Sanders, Jr. Reston, Va.: Reston Publishing, 1977. 299p. illus. index. LC 76-30480. ISBN 0-879-09223-8.

Provides electronics technicians and students with a practical guide to trouble localization in a broad range of electronic equipment. Areas covered are performance verification and calibration, DC voltage measuring, resistance, DC and AC current measuring methods and procedures, radio, HI FI, and television troubleshooting.

383.   **Electronic Test Equipment.** By T. J. Byers. New York: McGraw-Hill, 1987. 320p. illus. index. ISBN 0-07-009522-1. $39.95.

Provides thorough coverage of the theory and operation of a wide range of electronic test equipment from simple voltmeters to specialized analyzers. The text is organized by electronic function and includes chapters on basic meters, signal generators, tracers, counters, and oscilloscopes.

384.   **Electronic Test Equipment, Operation & Applications.** Edited by A. M. Rudkin. London: Granada, 1981. 316p. illus. index. ISBN 0-246-11478-9. $49.00.

A discussion of the general operations and applications of frequently used electronic test equipment. Sources of error are given and ways to maximize the equipment are covered in some detail. Coverage includes low-frequency oscillators, sweep generators, AF and RF power meters, modulation meters, oscilloscopes, and component bridges.

385.   **Handbook of Advanced Solid-State Troubleshooting.** By Miles Ritter-Sanders, Jr. Reston, Va.: Reston Publishing, 1977. 255p. illus. index. LC 77-23190. ISBN 0-87909-321-8. $21.95.

Explains and illustrates state-of-the-art analytical techniques, tests, measurements, and the practical troubleshooting of solid-state devices, circuits, and systems. Coverage includes semiconductor types, fundamental troubleshooting procedures, audio amplifier testing, solid-state stereo systems, radio receivers, television troubleshooting, digital troubleshooting methods, and advanced semiconductor tests. Instruments covered include multimeters, hi-lo transistor voltmeters, digital frequency counters, oscilloscopes, logic probes, pulse generators, signal tracer, semiconductor testors, etc.

386.   **Handbook of Electronic Test Procedures.** John D. Lenk. Englewood Cliffs, N.J.: Prentice-Hall, 1982. 320p. illus. index. LC 81-5218. ISBN 0-13-377457-0. $24.95.

This book is devoted to step-by-step instructions and procedures for testing electronic devices, components, and circuits. Each test is described along with how to perform the test, what is being tested, and why it is being tested. Three levels of tests are discussed: those using elementary equipment like meters; those using more advanced equipment, like oscilloscopes; and those using specialized equipment and tests like curve tracers. Each discussion includes examples of what type of readout or display to expect when testing good or bad components.

387.   **Handbook of Microwave Testing.** By Thomas S. Laverghetta. Dedham, Mass.: Artech House, 1981. 350p. illus. index. LC 81-67941. ISBN 0-89006-070-3. $40.00.

A popular guide to microwave measuring and testing techniques, this handbook includes equipment types, methods for measuring peak microwave power, noise measurement, receiver measurements, and procedures for measuring active devices as well as a detailed discussion of the IEEE 488 standard. Contents consist of microwave test equipment, power measurements, spectrum analyzer measurements, automatic testing, and miscellaneous measurements.

388.   **Handbook of Oscilloscopes: Theory and Application.** Rev. ed. By John D. Lenk. Englewood Cliffs, N.J.: Prentice-Hall, 1982. 320p. illus. index. LC 81-10680. ISBN 0-13-380576-x. $24.95.

Intended to supplement the operating instructions for major types of oscilloscopes, this cookbook is intended for beginners but can also be used as a guidebook for experienced users. Coverage includes applications data on the uses of oscilloscopes.

389. **Handbook of Practical Microcomputer Troubleshooting.** By John D. Lenk. Reston, Va.: Reston Publishing, 1979. 389p. illus. index. LC 79-1357. ISBN 0-8359-2757-1. $24.95.

The stated purpose of this book is to provide a simplified system of troubleshooting for the many microcomputers in use. Because of the large numbers of microcomputers and peripherals, this book concentrates on basic approaches which can be used for microprocessor/stored program systems. The author has written a number of best-selling electronics books.

390. **Interference Handbook.** Edited by William R. Nelson. Wilton, Conn.: Radio Publications, Inc., 1981. 247p. illus. index. LC 81-51709. ISBN 0-933616-01-5. $9.95.

Designed to help users locate and resolve all types of interference problems, this handbook describes sources of interference as well as methods for locating and suppressing problems. Suppression circuits are discussed in detail.

391. **Introduction to Electrical Instrumentation and Measurement Systems.** 2nd ed. By B. A. Gregory. New York: Halsted Press, 1981. 446p. illus. index. LC 80-22869. ISBN 0-470-27092-6. $29.95.

The goal of this book is to help the engineer/instrument user select the right type of instrument for his or her application by analyzing the performance of similar instruments from various manufacturers. References are found at the end of each section. The instruments covered include analog instruments like oscilloscopes, AC-DC potentiometers, AC-DC bridges, and transducers and signal conditioning instruments.

392. **New Digital Troubleshooting Techniques: A Complete Illustrated Guide.** By Robert G. Middleton. Englewood Cliffs, N.J.: Prentice-Hall, 1984. 280p. illus. index. LC 83-16157. ISBN 0-13-612275-2. $24.95.

Chapter headings include digital troubleshooting, latch troubleshooting, counter, encoders and decoders, multiplexers and demultiplexers, and generalized troubleshooting. Both digital test procedures and trouble analysis are discussed. Unique troubleshooting techniques like latch troubleshooting with a charge-storage probe and ohmmeter tests of a NOR gate are included.

393. **New Ways to Use Test Meters: A Modern Guide to Electronic Servicing.** By Robert G. Middleton. Englewood Cliffs, N.J.: Prentice-Hall, 1983. 256p. illus. index. LC 83-3348. ISBN 0-13-616169-3. $21.95.

Details both analog and digital techniques. Illustrated examples of specific guidelines are provided along with some calculator-aided troubleshooting methods.

394. **Practical Handbook of Solid State Troubleshooting.** By Robert C. Genn. West Nyack, N.Y.: Parker Publishing, Co., 1984. 256p. illus. index. LC 80-23897. ISBN 0-13-691295-8. $12.95.

Covers troubleshooting of solid-state audio amplifiers, servicing integrated circuits, regulated power supplies, television circuits, oscillators and waveshaping circuits, digital circuits, and test equipment.

395. **Tested Electronic Troubleshooting Methods.** 2nd ed. By Walter H. Buchsbaum. Englewood Cliffs, N.J.: Prentice-Hall, 1982. 272p. illus. index. LC 82-13167. ISBN 0-13-906966-6. $19.95.

Provides over 100 diagrams of circuits and components showing how to repair color televisions, stereos, transistor circuits, analog ICs, digital ICs, microprocessor controlled devices, and mini or microcomputers. Each chapter gives a clear summary of the method described along with relevant safety warnings such as where high voltages occur. The discussion of safety issues increases the usefulness of this book. The author has written a number of books in electronics.

396. **Troubleshooting and Repairing Electronic Test Equipment.** 2nd ed. By Mannie Horowitz. Blue Ridge Summit, Pa.: Tab Books, 1986. illus. index. LC 86-2345. ISBN 0-8306-0663-7. $24.95.

The previous edition of this book was very popular with technicians as well as engineering students. The second edition has been substantially enlarged to include all major modern test equipment. Special emphasis is given to self-testing equipment.

397. **Troubleshooting & Repairing Personal Computers.** By Art Margolis. Blue Ridge Summit, Pa.: Tab Books, 1983. 320p. illus. index. LC 82-19342. ISBN 0-8306-1539-3. $14.50.

A treatment of the new and expanding field of microcomputer repair. The first eight chapters cover chip changing. Chapter 9 and 10 discuss digital electronics while chapters 11 through 21 detail actual technical repairs. There are numerous diagrams and schematics.

398. **Understanding and Troubleshooting the Microprocessor.** By James W. Coffron. Englewood Cliffs, N.J.: Prentice-Hall, 1980. 338p. illus. index. LC 79-20950. ISBN 0-13-936625-3. $25.95.

Covers semiconductor memories, construction of a keyboard, 8080 circuit as a CPU, microprocessor I/O, and taking advantage of LSI. The two appendixes include data sheets, parts lists, and schematics. The parts lists for keyboards and CPUs make this book particularly useful.

PART 4
Journals and
Newsletters

# 15 Association, Trade, and Scholarly Journals

Since the inception of the first journal, *Journal Des Scavans* in France over 300 years ago, journal publishing has grown exponentially. Journals, periodicals, and magazines are now published in every discipline. Like journals in all fields, those that specialize in the areas of electrical, electronics, and computer engineering serve two basic purposes: as a historical record of activity and as a method for exchanging information.

For most of the technical journals listed in this section, the primary goal is the dissemination of information while the secondary goal is to provide a record of scientific activity. To serve these objectives, journals may be published by professional societies, trade associations, commercial publishers, universities, companies, or governmental agencies. They may cover alone or in any combination scholarly research, industry news, new products, public relations, or research by a specific agency or institution.

The following list consists of examples of journals found in electrical and electronics engineering. Periodicity, ISSN number, and abstracting or indexing information is given where known. Subscription costs are provided but are subject to change. A number of journals are "free to qualified" with those who qualify usually meaning design engineers, procurement officers, and others who are authorized to buy equipment or software. Some indication of the contents is also noted. For this section the word "advertisements" means that a substantial part of the journal is made up of ads which are scattered throughout the text. Some scholarly publishers promote related books or journals at the end of selected issues. These are not considered to be advertisements. The terms "job board or job listings" means that the journal includes advertisements by employers of available positions. Product literature means that a mechanism, usually a postcard, is available for obtaining product literature on products either reviewed or advertised in the journal.

Abbreviations used for index sources:

| | | |
|---|---|---|
| ASTI | = | *Applied Science and Technology Index* |
| CA | = | *Chemical Abstracts* |
| Comp Abst | = | *Computer Abstracts* |
| Comp Rev | = | *Computing Reviews* |
| Curr Cont | = | *Current Contents* |
| EI | = | *Engineering Index* |
| Sci Abst | = | *Computers and Control, Electrical and Electronics* and *Physics Abstracts* |
| Sci Cit | = | *Science Citation Index* |

399. **Canadian Electrical Engineering Journal.** Vol. 1- . Montreal, Quebec: Canadian Society for Electrical Engineers, 1976- . quarterly. ISSN 0700-9216. $30.00/yr.
Indexed: EI, Sci Abst.
Text and summaries in English and French.

400. **Canadian Electronic Engineering.** Vol. 1- . Toronto, Ontario: Maclean-Hunter Ltd., 1957- . 10 times/yr. ISSN 0008-3461. $45.00/yr.
Indexed: Sci Abst.
Text and summaries in English and French.

401. **Chilton's Electronic Component News.** Vol. 1- . Radnor, Pa.: Chilton, Co., 1957- . monthly. ISSN 0193-614x. $36.00/yr. (Formerly *Electronic Component News*).
Includes: advertisements, product literature.

402. **Circuit.** Toronto, Ontario: Electrical and Electronic Manufacturers Association of Canada. bimonthly. ISSN 0381-1905. $35.00/yr.

403. **Circuit World.** Vol. 1- . Ayr, Great Britain: Wela Publications, 1973- . quarterly. $38.50/yr.
Indexed: Sci Abst.

404. **Circuits Manufacturing.** Vol. 1- . Boston: Benwill Publishing, 1961- . monthly. ISSN 0009-7306. free to qualified, others $20.00/yr. (Formerly *Electronics Production*).
Indexed: Sci Abst.
Includes: advertisements, book reviews, annual index.

405. **Computer.** Vol. 1- . Piscataway, N.J.: IEEE Press, 1966- . monthly. ISSN 0018-9162. $90.00/yr. (Formerly *IEEE Computer Group News*).
Indexed: EI, Sci Abst, Comp Rev.
Includes: advertisements, book reviews, new book lists, job listings.

406. **Computer Design**. Vol. 1- . Littleton, Mass.: PennWell Publishing, 1962- . monthly. ISSN 0010-4566. free to qualified, others $50.00/yr.
Indexed: EI, Sci Abst, ASTI, Curr Cont.

407. **Computers and Electrical Engineering**. Vol. 1- . New York: Pergamon Press, 1973- . quarterly. ISSN 0045-7906. $190.00/yr.
Indexed: EI, Sci Abst.

408. **Computers and Electronics**. Vol. 1- . New York: Ziff-Davis, 1954- . monthly. ISSN 0745-1458. $15.95/yr. (Formerly *Popular Electronics*).
Indexed: *Readers Guide*.
Includes: advertisements, book reviews, product literature, semi-annual index.

409. **CSSP: Circuits Systems and Signal Processing**. Vol. 1- . Boston: Birkhauser, 1980- . quarterly. ISSN 0278-081x. $98.00/yr.
Indexed: Sci Abst.

410. **Defense Electronics**. Vol. 1- . Palo Alto, Calif.: E. W. Communications, 1969- . ISSN 0-278-3479. $28.00/yr. (Formerly *Electronic Warfare/Defense*).
Indexed: Sci Abst.
Includes: advertisements, a job board, product literature.

411. **Design and Test of Computers**. Vol. 1- . Piscataway, N.J.: IEEE Press, 1984- . quarterly. ISSN 0740-7475. $60.00/yr.

412. **Digital Processes: An International Journal on the Theory and Design of Digital Systems**. Vol. 1- . New York: Crane Russak, 1980- . quarterly. ISSN 0196-599x. $80.00/yr.
Indexed: Comp Abst, EI, Sci Abst, Comp Rev.

413. **EC&M: Electrical Construction and Maintenance**. Vol. 1- . New York: McGraw-Hill, 1900- . monthly. ISSN 0013-4260. $18.00/yr.
Indexed: ASTI.
Includes: Advertisements, product or equipment notices, and requests for bids.

414. **EDN/Electrical Design News**. Vol. 1- . Boston: Cahners Publishing, 1956- . 30 times/yr. ISSN 0012-7515. free to qualified, others $74.00/yr.
Indexed: EI, Sci Abst.
Includes: editorials, product reviews, advertisements, index to editorials.

415. **E E: Evaluation Engineering**. Vol. 1- . Highland Park, Ill.: A. Verner Nelson Associates, 1962- . 10 times/yr. ISSN 0149-0370. free to qualified, others $54.00/yr. (Formerly *Evaluation Engineering*).
Indexed: Sci Abst.
Includes: advertisements, product literature.

416. **Electric Light and Power: The News Magazine of Electric Utility Management and Technology**. Vol. 1- . Barrington, Ill.: Technical Publishing, Dun & Bradstreet, 1922- . monthly. ISSN 0013-4120. $38.00/yr.
Includes: advertisements.

417. **Electric Power Systems Research: An International Journal Devoted to Research Applications in Generation, Transmission, Distribution and Utilization of Electric Power.** Vol. 1- . New York: Elsevier Science Publication, 1978- . quarterly. ISSN 0378-7796. $112.00/yr.
    Indexed: CA, Sci Abst, Curr Cont.

418. **Electric Technology USSR.** Vol. 1- . Elmsford, N.Y.: Pergamon Press: 1958- . quarterly. ISSN 0013-4155. $290.00/yr.
    Indexed: CA, EI, Sci Abst, Curr Cont.
    English translation of Elektrichestvo.

419. **Electrical Engineer: Power Generation, Electricity, Transmissions and Utilizations.** Vol. 1- . Chippendale, Australia: Thomson Publications, 1924- . monthly. ISSN 0013-4309. $30.00/yr.
    Indexed: CA, EI, Sci Abst.
    Includes: advertisements, books reviews, index.

420. **Electrical Engineering in Japan: Scripta Electronica Japonica.** Vol. 1- . Silver Spring, Md.: Scripta Publishing, 1963- . bimonthly. ISSN 0424-7760. $299.00/yr.
    Indexed: EI, Sci Abst.
    Includes: editorials, patent lists, index.
    English translation.

421. **Electrical Power Engineer.** Vol. 1- . Surrey, Great Britain: Electrical Power Engineers Association, 1919- . monthly. ISSN 0013-43-76. $15.00/yr.

422. **Electrical Review.** Vol. 1- . Surrey, Great Britain: Quadrant House, IPC, 1872- . 5 times/yr. ISSN 0013-4384. $85.00/yr.
    Indexed: Sci Abst, CA.
    Includes: book reviews, a job board, patent lists, product literature, index.
    Covers European aspects of electrical engineering.

423. **Electrical Times.** Vol. 1- . Surrey, Great Britain: Quadrant House, 1891- . 5 times/yr. ISSN 0013-4414. $47.50/yr.
    Indexed: Sci Abst.
    Includes: advertisements, book reviews, a job board, product literature, semi-annual index.

424. **Electrical World.** Vol. 1- . New York: McGraw-Hill, 1896- . monthly. ISSN 0013-4457. $11.00/yr to qualified, others $38.00/yr.
    Indexed: ASTI, EI, CA, Sci Abst.
    Includes: advertisements, book reviews, job listings, product literature, index.

425. **Electri-Onics.** Vol. 1- . Libertyville, Ill.: Lake Publishing, 1955- . monthly. ISSN 0020-4544. $48.00/yr. (Formerly *Insulation/Circuits*).
    Indexed: EI, CA.
    Includes: advertisements, book reviews, product literature.

426. **Electrocomponent Science and Technology.** Vol. 1- . London: Gordon and Breach Science Publishing, 1974- . monthly. ISSN 0305-4643. $224.00/yr.
Indexed: Sci Abst.
Includes: book reviews, annual index.

427. **Electronic Design.** Vol. 1- . Rochelle Park, N.J.: Hayden Publishing, 1953- . semimonthly. ISSN 0013-4872. $40.00/yr.
Indexed: ASTI, EI, Sci Abst.
Includes: advertisements, book reviews, patents, product literature.

428. **Electronic Engineering.** Vol. 1- . London: Morgan Grampian, 1928- . monthly. ISSN 0013-49-2. free to qualified in United Kingdom, others $47.00/yr.
Indexed: EI, CA, Sci Abst, Curr Cont.
Includes: advertisements, product literature, annual index.

429. **Electronic Engineering Times.** Vol. 1- . Manhasset, N.Y.: CMP Publications, 1972- . semimonthly. ISSN 0192-1541. $15.00/yr.
Includes: advertisements, new products.

430. **Electronic Packaging and Production.** Vol. 1- . Denver: Cahners Publishing, 1961- . monthly. ISSN 0013-49-45. $55.00/yr.
Indexed: EI, Sci Abst.
Includes: advertisements, job listings, product literature, index.

431. **Electronic Product Design.** Vol. 1- . Bromley, Great Britain: Walton Hours, 1980- . monthly. ISSN 0360-1307. $50.00/yr.
Indexed: Sci Abst.
Includes: advertisements, book reviews.

432. **Electronic Products Magazine.** Vol. 1- . Garden City, N.J.: Hearst Business Communications, 1958- . 15 times/yr. ISSN 0013-4953. free to qualified, others $30.00/yr.
Indexed: EI, Sci Abst.
Includes: advertisements, book reviews, job listings, product literature.

433. **Electronics and Communications: The Engineering Journal of the Canadian Electronics Industry.** Vol. 1- . Don Mills, Ontario: Southam Communications, 1953- . 7 times/yr. ISSN 0013-5100. $25.50/yr.
Indexed: Sci Abst.
Includes: advertisements, book reviews, product literature, index.

434. **Electronics and Communications in Japan.** Vol. 1- . New York: John Wiley, 1985- . quarterly. ISSN 0424-8368. $299.00/yr.
Indexed: EI, Sci Abst.
Includes: patents, index.
A two-part publication consisting of communications and electronics. Both are translated from *Transactions of Institute of Electronics and Communications Engineers of Japan.*

435. **Electronics and Power.** Vol. 1- . London: IEE, 1955- . 11 times/yr. ISSN 0013-5127. $146.00/yr.
> Indexed: CA, EI, Sci Abst.
> Includes: book reviews, ads for IEE materials and conferences, calls for papers, book reviews, index.

436. **Electronics Industry.** Vol. 1- . New Malden, Great Britain: E S Publications, 1959- . monthly. ISSN 0307-2401. free to qualified in United Kingdom. (Formerly *Electronics Components*).
> Indexed: Sci Abst.
> Includes: advertisements, book reviews, index.

437. **Electronics Letters.** Vol. 1- . London: IEE, 1975- . bimonthly. ISSN 0013-5194. $366.00/yr.
> Indexed: EI, Sci Abst.
> Includes: index.
> A rapid publication journal.

438. **Electronics Week.** Vol. 1- . New York: McGraw-Hill, 1930- . weekly. ISSN 0748-3252. $52.00/yr. (Formerly *Electronics*).
> Indexed: ASTI, EI, CA, Curr Cont.
> Includes: advertisements, book reviews, product literature, index.

439. **Electronics Weekly.** Vol. 1- . West Sussex, Great Britain: IPC Electronic Press, 1960- . weekly. ISSN 0013-5224. free to qualified, others $40.00/yr.
> Includes: advertisements, job listings, product literature, book reviews.

440. **Electro-Optics.** Vol. 1- . Littleton, Mass.: PennWell Publishers, 1985- . 20 times/yr. ISSN 8756-7180. $200.00/yr. (Formerly *Electro-Optical Systems Design*).
> Indexed: EI, Curr Cont.
> Includes: marketing and financial data.

441. **Electrotechnology.** Vol. 1- . London: IEE, 1965- . quarterly. ISSN 0306-8552. $12.00/yr. (Formerly *Electrical and Electronics Incorporated Engineers*).
> Indexed: Sci Abst.

442. **Hybrid Circuit Technology.** Vol. 1- . Libertyville, Ill.: Lake Publishing, 1984- . bimonthly. ISSN 0747-1599. $60.00/yr.

443. **IEE Proceedings.** London: IEE, 1980- .

> **IEE Proceedings Part B: Electric Power Applications**. Vol. 1- . London: IEE, 1980- . bimonthly. ISSN 0143-7038. $646.15 for all 10 parts.
>> Indexed: Sci Abst.
>> Includes: book reviews, index.

> **IEE Proceedings Part C: Generation, Transmission and Distribution.** Vol. 1- . London: IEE, 1980- . bimonthly. ISSN 0143-7046.
>> Indexed: Sci Abst.
>> Includes: book reviews, index.

**IEE Proceedings Part D: Control Theory and Applications.** Vol. 1- . London, IEE, 1980- . bimonthly. ISSN 0143-7054. $346.15 for parts B, C, D.
    Indexed: Sci Abst.
    Includes: book reviews, index.

**IEE Proceedings Part F: Communications, Radar and Signal Processing.** Vol. 1- . London: IEE, 1980- . monthly. ISSN 0143-7070.
    Indexed: Sci Abst.
    Includes: book reviews, index.

**IEE Proceedings Part G: Electronic Circuits and Systems.** Vol. 1- . London: IEE, 1980- . bimonthly. ISSN 0413-7089.
    Indexed: Sci Abst.
    Includes: book reviews, index.

**IEE Proceedings Part H: Microwaves, Optics and Antennas.** Vol. 1- . London: IEE, 1980- . bimonthly. ISSN 0143-7097.
    Indexed: Sci Abst.
    Includes: book reviews, index.

**IEE Proceedings Part I: Solid-State and Electron Devices.** Vol. 1- . London: IEE, 1980- . bimonthly. ISSN 0143-7100.
    Indexed: Sci Abst.
    Includes: book reviews, index.

444.   **IEEE Publications.** Piscataway, N.J.: IEEE.

**IEEE Circuits and Devices Magazine.** Vol. 1- . Piscataway, N.J.: IEEE Press, 1985- . bimonthly. ISSN 8755-3996. $72.00/yr.
    Indexed: Sci Abst.
    Includes: new books list, book reviews.

**IEEE Control Systems Magazine.** Vol. 1- . Piscataway, N.J.: IEEE Press, 1981- . quarterly. ISSN 0272-01708. $53.00/yr.
    Indexed: EI, Sci Abst.
    Includes: book reviews, lists of other IEEE publications of interest, index.

**IEEE Journal of Quantum Electronics.** Vol. 1- . Piscataway, N.J.: IEEE Press, 1965- . monthly. ISSN 0018-9197. $180.00/yr.
    Indexed: CA, EI, Sci Abst, Curr Cont.

**IEEE Journal of Solid-State Circuits.** Vol. 1- . Piscataway, N.J.: IEEE Press, 1966- . bimonthly. ISSN 0018-9200. $113.00/yr.
    Indexed: EI, Sci Abst.

**IEEE Power Engineering Review.** Vol. 1- . Piscataway, N.J.: IEEE Press, 1981- . monthly. ISSN 0272-1724. $65.00/yr.
    Indexed: Sci Abst.
    Includes: listings of IEEE conferences, index.

**IEEE Proceedings.** Vol. 1- . Piscataway, N.J.: IEEE Press, 1913- . monthly. ISSN 0018-9219. $140.00/yr.
Indexed: EI, Sci Abst.

**IEEE Spectrum.** Vol. 1- . Piscataway, N.J.: IEEE Press, 1954- . monthly. ISSN 0018-9235. $92.00/yr.
Indexed: ASTI, EI, Sci Cit.
Includes: book reviews, new book lists, job listings, product literature, calls for papers, index.

**IEEE Transactions on Acoustics, Speech and Signal Processing.** Vol. 1- . Piscataway, N.J.: IEEE Press, 1951- . bimonthly. ISSN 0096-3518. $129.00/yr.
Indexed: ASTI, CA, EI, Sci Abst.

**IEEE Transactions on Aerospace and Electronic Systems.** Vol. 1- . Piscataway, N.J.: IEEE Press, 1965- . monthly. ISSN 0018-9251. $114.00/yr.
Indexed: ASTI, EI, Sci Abst.

**IEEE Transactions on Antennas and Propagation.** Vol. 1- . Piscataway, N.J.: IEEE Press, 1952- . monthly. ISSN 0018-926x. $140.00/yr.
Indexed: ASTI, EI, Sci Abst.

**IEEE Transactions on Automatic Control.** Vol. 1- . Piscataway, N.J.: IEEE Press, 1956- . monthly. ISSN 0018-9286. $140.00/yr.
Indexed: ASTI, EI, Sci Abst.

**IEEE Transactions on Circuits and Systems.** Vol. 1- . Piscataway, N.J.: IEEE Press, 1952- . monthly. ISSN 0098-4094. $120.00/yr.
Indexed: ASTI, CA, EI, Sci Abst.

**IEEE Transactions on Components, Hybrids and Manufacturing Technology.** Vol. 1- . Piscataway, N.J.: IEEE Press, 1978- . quarterly. ISSN 0148-6411. $88.00/yr.
Indexed: ASTI, CA, EI, Sci Abst.

**IEEE Transactions on Computer-Aided Design of Integrated Circuits.** Vol. 1- . Piscataway, N.J.: IEEE Press, 1982- . quarterly. ISSN 0278-0070. $72.00/yr.
Indexed: Sci Abst.

**IEEE Transactions on Consumer Electronics.** Vol. 1- . Piscataway, N.J.: IEEE Press, 1952- . quarterly. ISSN 0098-3063. $72.00/yr.
Indexed: ASTI, CA, EI, Sci Abst.

**IEEE Transactions on Electrical Insulation.** Vol. 1- . Piscataway, N.J.: IEEE Press, 1965- . bimonthly. ISSN 0018-9367. $88.00/yr.
Indexed: ASTI, CA, EI, Sci Abst.

**IEEE Transactions on Electron Devices.** Vol. 1- . Piscataway, N.J.: IEEE Press, 1952- . monthly. ISSN 0018-9383. $175.00/yr.
Indexed: ASTI, CA, EI, Sci Abst.

**IEEE Transactions on Industrial Electronics.** Vol. 1- . Piscataway, N.J.: IEEE Press, 1953- . quarterly. ISSN 0278-0046. $72.00/yr.
Indexed: ASTI, CA, EI, Sci Abst.

**IEEE Transactions on Microwave Theory and Techniques.** Vol. 1- . Piscataway, N.J.: IEEE Press, 1953- . monthly. ISSN 0018-9480. $126.00/yr.
Indexed: ASTI, CA, EI, Sci Abst.

**IEEE Transactions on Power Apparatus and Systems.** Vol. 1- . Piscataway, N.J.: IEEE Press, 1884- . monthly. ISSN 0018-9510. $200.00/yr.
Indexed: ASTI, CA, EI, Sci Abst.
Includes: papers from the summer and winter meetings of the IEEE Power Engineering Society.

445. **Integrated Circuits International.** Vol. 1- . Oxford: Elsevier International, 1976- . monthly. ISSN 0263-6522. $240.00/yr.
Indexed: Sci Abst.

446. **Integrated Circuits Magazine.** Vol. 1- . Garden City, N.J.: Hearst Publications, 1975- . monthly. free to qualified. (Formerly *IC Master/IC Update*).

447. **International Journal for Hybrid Microelectronics.** Vol. 1- . Silver Spring, Md.: International Society for Hybrid Microelectronics, 1978- . semi-annual. ISSN 0277-8270. $63.00/yr.
Indexed: CA, Sci Abst.

448. **International Journal of Circuit Theory and Applications.** Vol. 1- . New York: John Wiley, 1973- . quarterly. ISSN 0098-9886. $320.00/yr.
Indexed: EI, Sci Abst, Curr Cont.
Includes: book reviews, index.

449. **International Journal of Electronics.** Vol. 1- . Basingstoke, Great Britain: Taylor and Francis, 1965- . monthly. ISSN 0020-7217. $519.00/yr.
Indexed: EI, Sci Abst, Curr Cont.
Includes: book reviews.

450. **Japan Electronics Today News.** Vol. 1- . Luton, Great Britain: Benn Electronics Publications, 1981- . bimonthly. ISSN 0261-3506. $495.00/yr.
Looseleaf covering new products from Japan.

451. **J E E: Journal of Electronic Engineering.** Vol. 1- . Tokyo, Japan: Dempa Publications, 1964- . monthly. ISSN 0385-4507. $45.00/yr.
Indexed: Sci Abst.
Includes: advertisements, product literature, index.
Text in English.

452. **J E I: Journal of the Electronics Industry.** Vol. 1- . Tokyo, Japan: Dempa Publications, 1954- . monthly. ISSN 0385-4515. $37.00/yr.
Indexed: Sci Abst.
Includes: advertisements, index.
Test in English.

453.  **Journal of Electronic Materials.** Vol. 1- . Warrendale, Pa.: AIME, 1972- . bimonthly. ISSN 0361-5235. $100.00/yr.
Indexed: CA, EI, Sci Abst, Curr Cont.

454.  **Journal of Power Sources.** Vol. 1- . Lausanne, Switzerland: Elsevier Sequoia, 1976- . monthly. ISSN 0378-7753. $338.95/yr.
Indexed: CA, Curr Cont, Sci Abst, Sci Cit.

455.  **Microelectronic Engineering.** Vol. 1- . New York: North-Holland, 1983- . quarterly. ISSN 0167-9317. $85.00/yr.
Indexed: Sci Abst.

456.  **Microelectronic Manufacturing and Testing.** Vol. 1- . Libertyville, Ill.: Lake Publishing, 1978- . monthly. ISSN 0161-7427. free to qualified, others $50.00/yr.
Includes: advertisements, product literature.

457.  **Microelectronics and Reliability.** Vol. 1- . New York: Pergamon Press, 1962- . bimonthly. ISSN 0026-2714. $230.00/yr.
Indexed: EI, Sci Abst.

458.  **Microelectronics Journal.** Vol. 1- . Luton, Great Britain: Benn Electronics Publications, 1967- . bimonthly. ISSN 0026-2692. $190.00/yr.
Indexed: EI, Sci Abst.
Includes: book reviews, index.

459.  **Microwave and RF.** Vol. 1- . Rochelle Park, N.J.: Hayden Publishing, 1962- . monthly. ISSN 0026-2919. free to qualified, others $30.00/yr. (Formerly *Microwaves*).
Indexed: Sci Abst, Curr Cont.
Includes: advertisements.

460.  **Microwave Journal.** Vol. 1- . Dedham, Mass.: Horizon House, 1958- . monthly. ISSN 0026-2897. free to qualified, others $36.00/yr.
Index: Sci Abst, Curr Cont.
Includes: advertisements, book reviews, abstracts of papers of interest, product literature, index.

461.  **Modern Electronics.** Vol. 1- . Port Washington, N.Y.: Cowan Publishing, 1978- . monthly. ISSN 0149-2357. $12.00/yr.
Includes: advertisements, book reviews, product literature.

462.  **MSN: Microwave Systems News.** Vol. 1- . Palo Alto, Calif.: E. W. Communications, 1971- . monthly. ISSN 0164-3371. $35.00/yr. (Formerly *Microwave Systems News*).
Indexed: Sci Abst.
Includes: advertisements, book reviews, patents, product literature, index.

463.  **Optical and Quantum Electronics.** Vol. 1- . London: Chapman and Hall, 1968- . bimonthly. ISSN 0306-8919. $190.00/yr. (Formerly *Opto-Electronics*).
Indexed: Sci Abst.

464.  **Printed Circuit Fabrication.** Vol. 1- . Alpharetta, Ga.: PMS Industries, 1978- . monthly. ISSN 0274-8096. free to qualified, others $25.00/yr.
Includes: advertisements.

465.  **Radio and Electronic Engineer.** Vol. 1- . London: Institution of Electronic and Radio Engineers, 1933- . monthly. ISSN 0033-7722. $115.00/yr.
Indexed: ASTI, EI, CA, Sci Abst, Sci Cit.

466.  **R.F. Design.** Vol. 1- . Englewood, Colo.: Cardiff Publishing, 1978- . monthly. ISSN 0163-321x. $15.00/yr.
Indexed: Sci Abst.

467.  **Scientia Electrica: Journal for Modern Problems of Theorical and Applied Electrical Engineering.** Vol. 1- . Basel, Switzerland: Swiss Federal Institute of Technology, 1955- . quarterly. ISSN 0036-8695. $45.00/yr.
Indexed: CA, EI, Sci Abst.
Includes: cumulated index.
Text in English, French, and German.

468.  **Semiconductor International: Processing, Assembly and Testing.** Vol. 1- . Denver, Colo.: Cahners Publishing, 1978- . monthly. ISSN 0163-3767. $55.00/yr.
Includes: advertisements.

469.  **Signal Processing.** Vol. 1- . Amsterdam: North-Holland, 1981- . 8 times/yr. ISSN 0165-1684. $160.00/yr.
Rapid publication journal.

470.  **Solid-State Electronics.** Vol. 1- . Elmsford, N.Y.: Pergamon Press, 1960- . monthly. ISSN 0038-1101. $300.00/yr.
Indexed: CA, Sci Abst, ASTI.
Includes: book reviews, lists of new books, index.

471.  **Solid-State Technology.** Vol. 1- . Port Washington, N.Y.: Cowan Publishing, 1958- . monthly. ISSN 0038-111x. $18.00/yr. (Formerly *Semiconductor Products and Solid State Technology*).
Indexed: CA, EI, Sci Abst, Curr Cont.
Includes: advertisements, book reviews, job listings, patents, index.

472.  **Soviet Electrical Engineering.** Vol. 1- . New York: Allerton, 1975- . bimonthly. ISSN 0038-5379. $330.00/yr.
Indexed: CA.
English translation of *Elektroteknika*.

473.  **Soviet Microelectronics.** Vol. 1- . New York: Consultants Bureau, 1972- . bimonthly. ISSN 0363-8529. $225.00/yr.
Indexed: Curr Cont.
English translation of *Mikroelectronika*.

474. **Telecommunications.** Vol. 1- . Dedham, Mass.: Horizon House, 1967. monthly. ISSN 0040-2494. free to qualified, others $36.00/yr.
  Includes: advertisements, job listings, product literature.

475. **Telecommunications and Radio Engineering.** Vol. 1- . Silver Spring, Md.: Scripta Publishing, 1963- . monthly. ISSN 0040-2508. $299.00/yr.
  Indexed: EI, CA, Sci Abst, Curr cont.
  English translation of *Elektrosvyaz and Radiotekhnika.*

476. **Telephone Engineer and Management.** Vol. 1- . Cleveland, Ohio: Harcourt Brace Jovanovich, 1909- . semimonthly. ISSN 0040-263x. $24.00/yr.
  Includes: advertisements, book reviews, product literature, job listings, conference listings.

477. **Telephony.** Vol. 1- . Chicago: Telephony Publishing, 1901- . weekly. ISSN 0040-2656. $30.00/yr.
  Includes: advertisements, book reviews, semi-annual index.

478. **VLSI Design.** Vol. 1- . Manhasset, N.Y.: CMP Publications, 1980- . monthly. ISSN 0279-2834. $19.95/yr. (Formerly *Lambda*).
  Includes: advertisements, book reviews, product literature.

479. **Volt.** Vol. 1- . Tokyo, Japan: NBS Publications, 1958- . monthly. ISSN 0042-8620. $35.00/yr.
  Includes: advertisements.
  Text in English.

# 16 House Technical Journals

Periodicals in the section are published by major companies in order to disseminate technical information to their engineers, to publicize achievement, and to alert associates to work being done. Although many of these companies also produce public relations journals, these are not included.

480. **AT&T Bell Laboratories Technical Journal.** Vol. 1- . Short Hills, N.J.: American Telephone and Telegraph Co., 1927- . 11 times/yr. ISSN 0005-8564.
    Indexed: CA, EI, Sci Abst, Curr Cont.
    Includes: lists of papers and patents by employees of Bell Laboratories.

481. **Brown Boveri Review.** English ed. Vol. 1- . Baden, Switzerland: Brown Boveri and Co., 1954- . monthly. ISSN 0007-2486. $83.00/yr.
    Indexed: CA, EI, Sci Abst, Curr Cont.
    A house publication of the large German electrical.

482. **Electrical Communication.** Vol. 1- . Harlow, Great Britain: ITT Corp., 1922- . quarterly. ISSN 0013-4252. $22.00/yr.
    Technical journal published in four languages, including English.

483. **EPRI Journal.** Vol. 1- . Palo Alto, Calif.: Electric Power Research Institute, 1976- . monthly. ISSN 0362-3416. free.
Indexed: Sci Abst.

484. **Fujitsu Scientific and Technical Journal.** Vol. 1- . Kanagawa-ken, Japan: Fujitsu Co., 1965- . quarterly. ISSN 0016-2523. $2.00/yr.
Indexed: CA, Curr Cont, EI, Sci Abst.
A house publication of Fujitsu, a Japanese manufacturing firm dealing primarily with electronic products.

485. **GEC Journal of Research.** Vol. 1- . Chelmsford, Great Britain: General Electric Co., 1983- . quarterly. ISSN 0264-9187. $30.00/yr.
Incorporates *Marconi Review.*

486. **Hewlett-Packard Journal.** Vol. 1- . Palo Alto, Calif.: Hewlett-Packard Co., 1949- . monthly. ISSN 0018-1153. free.
Indexed: Sci Abst.

487. **IBM Journal of Research and Development.** Vol. 1- . Armonk, N.Y.: IBM, 1957- . bimonthly. ISSN 0018-8646. $20.00/yr.
Indexed: Curr Cont, CA, Comp Abst, Comp Rev, EI, Sci Abst.
Includes: patents, other papers of employees.

488. **IBM Systems Journal.** Vol. 1- . Armonk, N.Y.: IBM, 1962- . quarterly. ISSN 0018-8670. $12.00/yr.
Indexed: Curr Cont, CA, Comp Abst, Comp Rev, Sci Abst, EI.

489. **IBM Technical Disclosure Bulletin.** Vol. 1- . Armonk, N.Y.: IBM, 1985- . monthly. ISSN 0018-8689. free to qualified.
Indexed: EI, Sci Abst.
Includes: patentable ideas which IBM does not wish to patent but does not want anyone else to be able to patent either.

490. **Mitsubishi Electric Advance.** Vol. 1- . Tokyo, Japan: Mitsubishi Electric Corp., 1978- . quarterly. ISSN 0386-5096. $25.00/yr.
Indexed: Sci Abst.

491. **National Technical Report.** Vol. 1- . Osaka, Japan: Matsushita Electric Co., 1955- . bimonthly. ISSN 0028-0291. $18.00/yr.
Indexed: CA, EI, Sci Abst.
Includes: product literature, index.
Consists of English summaries and illustration captions.

492. **NEC Research Development.** Vol. 1- . Tokyo, Japan: Nippon Electric Co., 1960-. quarterly. ISSN 0547-051x. exchange.
Indexed: Curr Cont, Sci Abst.

493. **Norelco Reporter.** Vol. 1- . Mahway, N.J.: Philips, 1954- . quarterly. ISSN 0029-1625. free to qualified.
Indexed: CA, Sci Abst.

494.   **RCA Review.** Vol. 1- . Princeton, N.J.: RCA, 1936- . quarterly. ISSN 0033-6831. $12.00/yr.
Indexed: Curr Cont, EI, Sci Abst.

495.   **Siemens Components.** Vol. 1- . Erlangen, West Germany: Siemens AG, 1906- . bimonthly. ISSN 0173-1726. free.
House publications of the German electrical firm.

496.   **Sumitomo Electric Technical Review.** Vol. 1- . Osaka, Japan: Sumitomo Electric Industries, 1962- . monthly. ISSN 0376-1207. exchange.
Indexed: Sci Abst.

497.   **Telesis.** Vol. 1- . Ottawa, Quebec: Bell-Northern Research Ltd., 1970- . quarterly. ISSN 0040-2710. free.
Indexed: Sci Abst.

498.   **Texas Instruments Engineering Journal.** Vol. 1- . Dallas, Tex.: Texas Instruments, 1984- . bimonthly. ISSN 0882-2557. $20.00/yr.
Includes: training opportunities, application notes, design news, abstracts of patents.

499.   **Toshiba Review.** Vol. 1- . Tokyo, Japan: Toshiba Corp., 1960- . bimonthly. ISSN 0040-9642. $36.00/yr.
Indexed: CA, Curr Cont, EI, Sci Abst.

500.   **Vitro Technical Journal.** Vol. 1- . Silver Spring, Md.: Vitro Corp., 1981- . free to qualified.
House publication of Vitro, a company specializing in electronics, communications, information, and energy.

# 17   Newsletters

Newsletters occupy a special place in the publishing world. They cannot be classified as books, journals, or technical reports because they do not publish the results of research or experimentation. Rather they are more like newspapers in that their goal is to disseminate information quickly. Usual coverage includes new product developments, sales figures, company information including research and development, and upper management and organizational news. Often newsletters include investigative reporting in an attempt to keep readers up to date on the activities of others. Such information is important to engineers, since so many are involved in activities that lead to developing, marketing, or selling products.

Newsletters can be divided into two types: commercial newsletter which focus on a particular industry or subject area, and those that are published by a specific company. The former are generally proprietary and have quite high subscription rates, while the latter are often inexpensive or free. Many proprietary newsletters have unique styles and editorial points of view and often rely on people in the industry to keep them informed of current news and even gossip.

Both types of newsletters are often emphemeral, are rarely indexed, and have poor bibliographic control. None of this negates their usefulness; it only makes them a bit more difficult to access. The effect of poor bibliographic control on the intended audience is quite minimal since most of these newsletters are directed at users who are highly motivated by business and professional considerations. Academic and public libraries are usually reluctant to purchase newsletters for a variety of reasons including cost, updating procedures, and the lack of indexing.

Newsletters come and go with amazing frequency. They may be produced by major publishers, individuals, or consulting firms. Unless issued by a major publisher, they are generally not available from vendors. The following selected list is as up to date and accurate as is possible with these important, but ephemeral, publications.

## Commercial Newsletters

501. **Advanced Electrical Information.** Walla Walla, Wash.: Donovan Gage. monthly. $45.00/yr.

502. **A/E Systems Report: Newsletter on Automation and Reprographics in Professional Design Firms.** Newington, Conn.: MRH Associates, Inc., 1977- . monthly. $72.00/yr.
   Examples of contents include tips for purchasing CADD systems, hardware/software news, and reprographics update.

503. **AI Trends: The Newsletter of the Artificial Intelligence Industry.** Scottsdale, Ariz.: DM Data, Inc., 1984- . monthly. $295.00/yr.
   Covers who is doing what, both people and companies.

504. **Anderson Report: Edited Exclusively for the Computer Graphics Industry.** Simi Valley, Calif.: Anderson Publishing Co., 1978- . monthly. $125.00/yr.
   Sections include trends, industry happenings, workstation to program robots, and inside report.

505. **Applied Artificial Intelligence Reporter.** Fort Lee, N.J.: ICS Research Institute, University of Miami, 1983- . monthly. $49.00/yr.
   Includes advertisements, book reviews, and news on AI.

506. **Breakthrough.** New York: Boardroom Reports Publishing. semimonthly. ISSN 0747-08000. $59.00/yr.
   Provides market analysis and forecasting data as well as pinpointing key innovators, developers, and sources of additional information. Covers fiber optics, ceramics, customized integrated circuits, and emerging technologies.

507. **Business Computer Report.** Surrey, Great Britain: Associated Publications for the Automation Industry, 1970- . monthly. $110.00.
   News and technical guidance for the professional community involved with automated information processing.

508. **Capital Investment Trends.** Bethesda, Md.: Economic Research Council. monthly. $96.00/yr.
   Surveys electronic industry investments in plants and new products.

509. **Chilton's Electronic Component News.** Radnor, Pa.: Chilton Co., 1957- . monthly. ISSN 0193-614x. $36.00/yr. (Formerly *Electronic Component News*).
   Product reviews and news on recently released products.

510.   **Circuit Alert.** Los Gatos, N. Mex.: Electronics Trade Press, 197?- . 25 times/yr. $100.00/yr.
Covers new circuits by category and gives a short synopsis with part number and manufacturer. Sections include A/D and D/A, power, telecommunications, optoelectronics, etc.

511.   **Competition and Strategy: A Monthly Interpretive News Report on Telecommunications.** Morristown, N.J.: Probe Research, Inc., 1979- . monthly. $140.00/yr.
Analysis of the telecommunications industry including individual companies.

512.   **Computer Age, Peripherals Digest: A Comprehensive Information Management Analysis of the Computer Peripherals Industry.** Annandale, Va.: EDP News Services, 1977- . monthly. $98.00/yr.
EDP also publishes *Computer Age, EDP Weekly* ($198.00/yr.).
Covers printers, modems, fax, and other computer peripherals.

513.   **Computer Aided Design Report.** San Diego, Calif.: CAD/CAM Publishing, 1980- . monthly. ISSN 0276-749x. $96.00/yr.
Continuing features include programming, company updates, literature, events, trade shows, and courses.

514.   **Conductor: Serving the Electronics Industry.** Alpharetta, Ga.: Conductor, 1981- . monthly. free.
Specializes in industry news and new products.

515.   **CRD Communications & Distributed Resources Report.** Framingham, Mass.: International Data Corp., 1978- . monthly. $395.00/yr.
Available electronically on NEWSNET. Specializes in news and business reports on the communications industry.

516.   **Data Channels.** Bethesda, Md.: Phillips Publishing, 1973- . weekly. $397.00/yr.
Covers the data communications market including local area networks, fiber optics, digital microwave, DTS/DEMS, videotext/teletext, and packet switching.

517.   **Datacom Reader Service.** Minneapolis, Minn.: Architecture Technology Corp., 197?- . monthly. $175.00/yr.
Abstracts new data communication product descriptions from key trade publications. Covers circuits, controllers, processors, electronic mail, facsimile, gateways, microwaves, modems, etc.

518.   **Dataquest.** San Jose, Calif.: Data Quest. Price varies from $7,000.00 to $15,000.00/yr.
An intelligence service which includes a number of newsletters. Domestic, European, Japanese, and Asian services are available. Examples of newsletters are: *Financial Services, Japanese Semiconductor Service, Small Computer Industry Service, Semiconductor Industry Service, Telecommunications Industry, CAD/CAM,* etc.

519. **Defense Economics Research Report.** New York: Data Resources Division of McGraw-Hill, 1980- . monthly. $495.00/yr.
Covers the defense industry including avionics and defense electronics.

520. **Defense Week.** Washington, D.C.: King Publishing Corp., 197?- . weekly. $595.00/yr.
Included because electronics plays an increasing role in weaponry.

521. **EDP: International Data Corporation's Newsletter for Executives Concerned with Electronic Data Processing.** Framingham, Mass.: International Data Corp. semimonthly. $365.00/yr.
Also known as "The Gray Sheet." Up-to-date reports on electronic data processing aimed at the corporate decision maker.

522. **EDP Japan Report.** Tokyo, Japan: International Data Corp. monthly. $20.00/yr.
Contains information collected about the Japanese computer industry and market.

523. **Electric Utility Week.** New York: McGraw-Hill, 1970- . weekly. ISSN 0046-1695. $825.00/yr. (Formerly *Electrical Week*).
Current information on the nation's utilities with special emphasis on new technology.

524. **Electronic Business Forecast.** San Jose, Calif.: Cahners Publishing, 1982- . 5 times/yr. $217.00/yr.
A business forecasting service in electronics.

525. **Electronic Market Trends.** Vol. 1- . Washington, D.C.: Electronic Industries Association, 1964- . monthly. $150.00/yr.
Analysis demand for future electronic and electrical products.

526. **Electronic News.** New York: Fairchild Publications, 1957- . weekly. $25.00/yr.
Electronics and computer markets are covered including government, military, aerospace, industrial, commercial, consumer, and international news.

527. **Electronic Warfare Digests.** Edited by Gerald Green. Annandale, Va.: Washington National News Reports. monthly. $90.00/yr.
Gives information on people, contracts, companies, the defense department, and the armed services. Discusses the latest procurement policies, contracts, and research and development information. Covers electronic countermeasures, electronic warfare, and electronic counter-countermeasures.

528. **Electronics Bulletin: What's New in Electronic Technology Today.** Ridgewood, N.J.: Media General Publication, 1985- . monthly. free to qualified.
Toll free telephone number is provided for requesting additional information from advertizers. Organized into product sections.

529. **EMMS: Electronic Mail & Message Systems.** Norwalk, Conn.: International Resource Development. bimonthly. ISSN 0163-9811. $235.00/yr.
Covers technology, users, products, and legislative trends in graphic and record communications.

530. **Fiber Optics and Communications Newsletter.** Boston: Information Gatekeepers, Inc. monthly. $215.00/yr.

Focuses on new developments in fiber optics and applications of this technology in any field.

531. **Fiber Optics Now: The Canstar Newsletter of Fiber Optics Technology.** Scarbourgh, Ontario: Canstar, 197?- . quarterly. $100.00/yr.

532. **Fiber Optics Patents.** Brookline, Mass.: Information Gatekeepers, Inc. monthly. $200.00/yr.

533. **Fiber Optics Weekly News Service.** Brookline, Mass.: Information Gatekeepers, Inc. monthly. $128.00/yr.

Newsletter covering the fiber optics industry, including users of fiber optic technology.

534. **GaAS News: Gallium Arsenide, III-V Newsletter.** Santa Clara, Calif.: 1985- . monthly. $450.00/yr.

Provides monthly capsule of the activities in the gallium arsenide and other compound semiconductor industries.

535. **Gartnergram.** Stamford, Conn.: Gartner Group. $295.00/yr.

Assessments by the Gartner Group of rumors circulating in the information industry. A credibility rating is given.

536. **High Tech Materials Alert.** Englewood, N.J.: Technical Insights, Inc., 1972- . monthly. ISSN 0741-0808. $237.00/yr.

Surveys international research and development in the manufacturing of new products and materials.

537. **High Technology Intelligence Report: Published in Silicon Valley from "Inside" Japan.** Sunnyvale, Calif.: High Technology Community Service, 1980- . published every 7 to 10 days. $180.00/yr.

All information contained in these reports is derived from Japanese-language publications. Covers semiconductors and ICs, computers and peripherals, software, fiber optics, etc.

538. **Hybrid Microelectronics Review.** Chalfont, Pa.: Matthew R. Romano, 1970- . monthly. $36.00/yr.

Attempts to keep readers up-to-date on the uses and new developments in hybrid circuits.

539. **Icecap Report: Covering the Semiconductory Industry.** Scottsdale, Ariz.: Integrated Circuit Engineering Corp. 3 times/yr. $395.00/yr.

Evaluative newsletter covering the world semiconductor industry.

540. **In-Stat Electronics Report: Industry Statistics and Analysis.** Scottsdale, Ariz.: Integrated Circuit Engineering. monthly. $475.00/yr.

Primary interest is the IC book, *Quarterly Trends for the USA*, which shows sales, percent change in sales, and charts giving booking/billing information.

541. **ISDN Newsletter.** Boston: Information Gatekeepers, Inc., 1985- . monthly. $195.00/yr.

Over 300 trade publications are covered with developments in the integrated services digital network industry reported. Also covers IDN and TDF news.

542. **Japan Electronics Today News.** Luton, Great Britain: Benn Electronics Publications, 1981- . semimonthly. ISSN 0261-3506. $495.00/yr.

Newsletter covering electronics and new electronic products in the Pacific Rim.

543. **Japan Semiconductor Quarterly.** Washington, D.C.: Electronic Industries Association of Japan, 1984- . quarterly. distributed as a public service by Gray and Company, Washington, D.C.

Aims to publish timely information and data of interest to users and manufacturers in both the United States and Japan. Emphasis is on the Japanese point of view.

544. **Japan Semiconductor Technology News.** Fort Lee, N.J.: Jack K. Burgess. bimonthly. ISSN 0286-8210.

545. **Lightwave.** Waltham, Mass.: Lightwave, 1984- . monthly. $40.00/yr.

Provides worldwide coverage of fiber optics, with each issue containing practical information about using fiber optics and advertisements from suppliers.

546. **Localnetter.** Minneapolis, Minn.: Architecture Technology Corp., 1981- . monthly. $300.00/yr.

Covers important developments in the field of local computer networks.

547. **Memory Update: The News Journal of Memory Technology.** Woodstock, Conn.: Spectrum Information Systems. semimonthly. $195.00/yr.

Provides information on new products and updates on the industry in the areas of rigid disc, semiconductor memory, and magnetic tape.

548. **Microelectronics News: With Manager's Casebook.** Pacific Grove, Calif.: Piquant. $250.00/yr.

A gossip sheet for the industry which covers employee benefits, likely merger candidates, information on actual mergers, who's dealing with the U.S. military, etc. Recent issues have focused on several companies alleged to be cheating the government.

549. **Microwave News.** New York: Louis Slesin, 1981- . 10 times/yr. ISSN 0275-6595. $200.00/yr.

Covers the microwave industry from an economic point of view.

550. **Monosson on DEC.** Boston: Monosson, 1982- . monthly. $395.00/yr.

An independent publication covering machines made by the Digital Equipment Company (DEC) as well as the equipment that is compatible with DEC.

551. **Nikkei High Tech Report.** Menlo Park, Calif.: NIRI, 1985- . semimonthly. $580.00/yr.

Produced by the Nikkei Industry Research Institute (NIRI), this newsletter is the English language version of *Nihon Keizai Shimbun*. It covers the latest developments, trends, new products, production techniques, marketing, and international trade and government policies in Japan in the areas of electronics, data processing, and telecommunications.

552. **Release 1.0.** New York: Rosen Research Inc., 1974- . 20 times/yr. $395.00/yr. (Formerly *Rosen Electronics Letter*).
Purports to give inside information on the electronics industry, including gossip, new products, company profits, etc.

553. **Robotics Report.** Annandale, Va.: Washington Capital News Department. monthly. $37.00/yr.
This newsletter deals with the growing field of robotics, including new products, research, and contracts.

554. **S. Klein Newsletter on Computer Graphics: News/Trends, Insights, Perspective and Opportunities.** Sudbury, Mass.: Technology and Business Communications Inc., 1979- . semimonthly. $155.00/yr.

555. **SCAN Newsletter.** Great Neck, N.Y.: SCAN, 1977- . monthly. $75.00/yr.
Attempts to keep manufacturers and users up to date on the advances and uses of bar code scanning.

556. **Scanning Electronic Technology (SET).** Fairfield, Conn.: Discovery Systems, Inc., 1982- . 3 times/yr. $85.00/yr.
Attempts to keep readers up-to-date on the developing uses of SET.

557. **Semiconductory Industry and Business Survey.** Sunnyvale, Calif.: HTE Management Resources, 1979- . 18 times/yr. ISSN 0730-1014. $400.00/yr.
See *SEMISTATS*.

558. **SEMISTATS: Semiconductor Industry Statistical Service.** Sunnyvale, Calif.: HTE Management Resources. $7,500.00/yr. Regular updates.
A complete subscription including telephone inquiry privileges. Provides company profiles, graphical analysis of industry statistics by major IC and discrete categories, industry and stock market charts of all publicly traded semiconductor firms. Can also be purchased in the following individual segments:
*Company Statistics.* Vol. 1. $1,950.00/yr. General information, technology and licensing agreements, plants and facilities, product family, and technology matix.
*IC-Statistics and Discrete-Statistics.* Vol. 2. $1,850.00/yr.
*Stock-Statistics and Financial Statistics.* Vol. 3. $750.00/yr.
*Personnel-Statistics.* Vol. 4.

559. **SIBS: Semiconductor Industry and Business Survey.** Sunnyvale, Calif.: HTE Management Resources, 1979- . 18 times/yr. ISSN 0730-1014. $495.00/yr.
News, facts, analysis, commentary, opinion, trends, views, and reports. Also available as part of the complete *SEMISTATS* subscription.

560. **Telecommunications Alert: A Digest of Telecommunications News.** New York: Management Telecommunications Publishing. 7 times/yr. ISSN 0742-5384. $149.00/yr.

Coverage includes market opportunities, new technologies, vendor hardware/software, legislation, system design and evaluation, and international development.

561. **Telecommunications Equipment News.** Arlington, Va.: Telecom Publishing, 1985- . 21 times/yr. $285.00/yr.

Product-oriented newsletter.

562. **Telecommunications Reports.** Washington, D.C.: Business Research Publications, 1934- . weekly. $294.00/yr.

Weekly news service which has covered the telecommunications, voice, record, and data service fields since 1934.

563. **Videospring.** Norwalk, Conn.: International Resource Development, 1979- . semimonthly. $180.00/yr.

Covers technology, and user, product, and legislative trends in home information systems and videotex and teletext systems and services.

564. **VLSI Manufacturing Newsletter.** San Jose, Calif.: VLSI Research. ISSN 8755-6219. $395.00/yr.

Aimed at manufacturers of VLSI equipment.

565. **VLSI Update: Personal Communique on International Trends and Developments.** Richardson, Tex.: Advance Associates, 1984- . ISSN 0270-85-7. $150.00/yr.

566. **Voicenews: Reporting on Speech Synthesis and Speech Recognition Technology.** Rockville, Md.: Stoneridge Technical Services, 1979- . 10 times/yr. $95.00/yr.

567. **Workstation Alert.** Boston: Management Roundtable, 1984- . $197.00/yr.

Monthly briefing service on engineering, scientific, and professional workstation applications and developments.

568. **World Electronic Developments.** New York: Prestwick Publications, 1981- . semiannual. ISSN 0740-3585. $165.00/yr.

Gives a worldwide overview and analysis of electronic development, with an analysis of the market and updates on new products.

# Company Newsletters

The newsletters in the following group are produced and edited by companies or laboratories. The intention is to alert users to new products, new research, and new applications of old products as well as to keep the company's products and services in the minds of users and prospective users. Like industry newsletters these are often ephemeral, without good bibliographic control, and are difficult to track down. The titles that follow were selected as examples; there has been no attempt to be exhaustive.

569.   **Communicator.** Santa Clara, Calif.: Calma Company, 1984- . bimonthly. free.
Newsletter of this subsidiary of General Electric.

570.   **Handshake: Tek Application Information.** Beaverton, Oreg.: Tektronix Corp.
quarterly. free.
Newsletter covering processing and instrument control.

571.   **Insight: News from Digital Equipment Corporation.** Northboro, Mass.: Digital
Equipment Corporation, 1980- . free.
Newsletter of the DEC, a large manufacturer of computers.

572.   **Motorola Semiconductor Data Update.** Phoenix, Ariz.: Motorola Semiconductor Products. bimonthly. $35.00/yr.
Consists of the data books, data sheets, applications notes, engineering bulletins,
product brochures, and other technical literature published by the Motorola Technical
Information Center. Also available on microfiche.

573.   **Signetics Now.** Sunnyvale, Calif.: Signetics Corp., 1983- . bimonthly. free.
Newsletter from this subsidiary of U.S. Phillips Corp.

574.   **TM Notes: Test and Measurement; A User Newsletter.** Beaverton, Oreg.:
Tektronix Corp., 1984- . quarterly. free.
Newsletter of a large manufacturer of electronic test equipment.

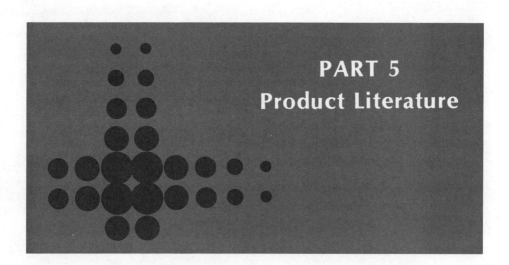

# PART 5
## Product Literature

# 18   Product Catalogs, Trade Directories, and Buyers' Guides

Service organizations, distributors, and trade associations are the most common sources of information on individual products, processes, materials, and services. Applications information is often combined with product information.

This literature is often ephemeral with the amount of information varying from one collection to another. Some publications merely include copies of advertisements while others provide data sheets and applications notes. Still others may consist entirely of lists of manufacturers by product. Catalogs are usually available from suppliers, as special issues of a technical journal, or may be purchased by subscription.

575.   **CAD/CAM Industry Directory.** Conroe, Tex.: Technical Database Corp., 1986- . ISSN 0736-1823. $25.00/yr.

Lists of software producers and hardware manufacturers in computer-aided design and manufacturing. Organized by broad subject categories, each entry includes a short description of the product with the company name, address, and phone number.

576.  **Circuit Manufacturers-Buyer's Guide.** Boston: Morgan Grampian, 1961- . annual. $30.00/yr. (Formerly *Circuits Manufacturers-Vendor Directory Issue*).

Covers 2,000 companies which produce artwork and provide etching, cleaning, coating, inspection, fabrication assembly, and soldering services for the circuit manufacturing industry.

577.  **Directory of Defense Electronic Products and Services (United States Suppliers).** 7th ed. By Electronic Industries Associations. New York: Information Clearing House, 1981. 210p. illus. index. ISBN 0-931634-06-7. $65.00.

Aims to be the single source of information on currently available U.S. defense electronic products and services. Product illustrations and technical specifications are provided in order to expedite procurement. A directory of suppliers and a glossary of terms are also included prior to the company name index.

578.  **EBG—Electronics Buyers Guide.** New York: McGraw-Hill, 19??. annual. $55.00/yr.

Consists of information from the primary manufacturers of electronic equipment in the United States and Europe. This directory of 4,000 products and services contains an advertisers' index, a list of catalogs available from the manufacturers, a directory of manufacturers and their sales offices, and lists of distributors. The company directory provides 5,000 telephone numbers of contact persons, as well as information on the number of engineers employed, the number of sales employees, sales volume, and geographic index of sales people.

579.  **EEM Electronic Engineers Master Catalog.** Garden City, N.Y.: Hearst Business Communications, 1958- . annual. $60.00.

Bound catalog of complete product data from the manufacturers of electronic and electromechanical components, systems, and equipment. Data is made up of pages duplicated from manufacturers catalogs. There are indexes by products, manufacturers, and trademarks. This is one of the best of the consolidated product guides.

580.  **Electrical and Electronics Trade Directory.** Stevenage, Great Britain: Peter Peregrinus; distr., Piscataway, N.J.: IEEE Press, 1883- . annual.

This directory provides information on companies, associations, and service bureaus involved in the electrical and electronics trade. Entries are organized into 3,750 product categories. The volume contains seven sections: a list of manufacturers, suppliers and servicing companies; manufacturers' representatives; wholesale distributors; associations, institutions, societies, and other organizations; electricity undertakings of the United Kingdom and Ireland; trade names; and products and materials. An alphabetical listing of subject headings and cross-references is also provided. Although international in coverage, the emphasis is on British firms.

581.  **Electronic Buyer's Handbook.** Manhasset, N.Y.: CMP Publications, 1976- . monthly. $28.95/vol.

A ten-volume set. Each volume, which covers a separate component, has a product, manufacturer, and distributor section.

582.  **Electronic Design's Gold Book.** New York: Hayden Publishing, 1973- . annual. $25.00/yr.

Published annually in four volumes as part of the magazine *Electronic Design*, the first volume covers amplifiers, crystals, crystal filters, oscillators, function modules, transformers, IC hybrids, etc. Each listing gives the product name, trade name, and manufacturer's profile. The other volumes include geographic listings and U.S. sales outlets, distributor profiles, and plant locations.

583. **Electronic Distributors Master Catalog.** Garden City, N.Y.: United Technical Publications, 1973- . annual. $45.00.
Basically contains copies of relevant manufacturers' catalog pages along with a directory of manufacturers and distributors.

584. **Electronic Miniatures: A Buyer's Guide.** By Susan Harris. Blue Ridge Summit, Pa.: Tab Books, 1982. 294p. illus. index. LC 82-5845. ISBN 0-8306-2476-7. $17.95.
This is a very selective buyers' guide, highlighting some of the thousands of products classed as electronic miniatures, which are the result of microelectronics, "the technology of constructing logic circuits and devices in extremely small packages." There are three background chapters covering the evolution of electronic miniatures. The major portion of the book, however, is dedicated to a buyers' guide of selected products, including television receivers, cassette recorders, calculators, pocket computers, clocks, microprocessors, and assorted gadgets. One shortcoming is the lack of a list of manufacturers and their addresses.

585. **Electro-Optical Systems Design-Vendor Selection.** Chicago: Cahners Publishing. annual. ISSN 0424-8457. $25.00/yr.
An alphabetical listing of 1,200 manufacturers and suppliers of equipment and materials used in the production and testing of electro-optical systems.

586. **Fiber Optics Handbook & Buying Guide.** 5th ed. Boston: Information Gatekeepers, Inc., 1983-1984. index. ISBN 0-686-32955-4. $50.00.
Includes some data on fiber optics, but is primarily a buyers' guide. Indexes are by broad subject and company name.

587. **Japan Electronics Almanac.** Tokyo, Japan: Dempa Publications, Inc., 1984- . annual. $150.00/yr. (Formerly *Japan Fact Book*).
Covers 60 leading Japanese electronics firms, with trade statistics, brand names, financial data, and usual company information.

588. **Laser Focus — Buyers' Guide.** Newton, Mass.: Advanced Technology Publications, 1964- . annual. ISSN 0075-8027. $35.00/yr.
Listings for 1,000 manufacturers and suppliers of products in the laser, fiber optic, and related industries. Entries include company name, address, telephone number, names of executives, number of employees, and lists of products/services. The text is arranged alphabetically with a separate product index.

589. **Microelectronic Manufacturing and Testing Desk Manual.** Libertyville, Ill.: Lake Publishing, 1974. irregular. ISSN 0161-7427. $25.00/yr.
This special issue of *Microelectronic Manufacturing and Testing* magazine describes materials, test instruments, parts, components, production equipment, and machines. Also lists accessories, suppliers, and services used to manufacture microelectronic, semiconductor, and solid-state products. Indexed by advertiser, supplier, and equipment.

590.   **Sensors and Transducer Directory.** Peterborough, N.H.: North American Technology, Inc., 1984- . annual. $25.00/yr.

Lists 315 manufacturers of sensors which are used in high technology industries. Each entry provides company addresses and brief descriptions of relevant products.

591.   **Solid-State Processing and Production: Worldwide Buyers' Guide and Directory.** Vol. 1- . New York: Dunn & Bradstreet Technical Publications, 1972- . annual. $60.00/yr.

Over 2,000 companies are represented in this source book for materials, equipment, and services used by the manufacturers of solid-state and related products, devices, and circuits. The guide, indexed by company and trade name, as well as by subject, covers a wide range of products from fume hoods to LED packages.

592.   **Sweet's Electrical Engineering Catalog File.** New York: McGraw-Hill, 1978- . annual. ISSN 0145-4897. $100.00/yr.

Like the other, more common Sweet's files, this consists of copies of pages from manufacturers' catalogs arranged in a classified order.

# 19  Company Data Books

Data books written and published by major electronics companies usually provide complete product data and device characteristics. While similar information can be found in the product catalog section of this book, there are a number of important differences between the two sources. Product catalogs are made up entirely of advertisements especially compiled by service organizations, trade magazines, or commercial publishers to help the user choose suitable products, equipment, and services and to obtain details about rival products. Data books are prepared by manufacturers to instruct the reader/user in the exploitation, maintenance, or use of a specific product. Because data books are often compilations of salesman's data sheets, they may not contain complete bibliographic information. (Data books are usually available from the salesman or are ordered directly from the company at a minimal charge.)

To help users who might like to order copies of the data books listed in this section, the address for each company is given followed by a selected list of available titles.

593.  Advanced Micro Devices, Inc.
      901 Thompson Pl.
      P.O. Box 453
      Sunnyvale, CA 94086
      (408) 732-2400 or (800) 538-8450

**Analog/Communications**
- Indexes/selection cross-references
- Package outlines/glossary

**Application Notes/Briefs/CMOS/Reliability Report**

**Bipolar Microprocessor Logic and Interface Numeric Device/Function Index**
- Numeric device/function index
- Design aids
- Systems considerations
- Am2960

**Bipolar PROM Cross-Reference Guide**

**Bipolar/MOS Memories Data Book**
- Indexes/selection guide/cross-reference
- Bipolar programmable memories

**MOS Microprocessors and Peripherals Indexes/Selection Guide, and Supplemental Data Sheets**
- Indexes, selection guide
- Z-Bus/6800 micro programmable bus translator
- Single chip microcomputers

**Programmable Array Logic and Supplemental Data**
- Instruction to programmable array logic
- Product specifications
- How to design with PALs
- Testing/programming/reliability information

594. Analog Devices
P.O. Box 280
Norwood, MA 02062
(617) 329-4700
**Application Notes**
**Integrated Circuits Data Acquisition Data Book**
**Modeules-subsystems Data Book**

595. Fairchild Camera and Instrument Corp.
Semiconductor Components Group
464 Ellis St.
P.O. Box 880A
Mountain View, CA 94042
(415) 962-5011
**Bipolar Memory Data Book**
**CCD Solid State Imaging Technology**
**CMOS Cross-Reference Guide**
**CMOS Data Book**
**Diode Data Book**
**Emitter Coupled Logic Data Book**
**Fast Fairchild Advanced Schottky TTL**
**F8 and F3870 Guide**
**F8 User Guide**
**F100K Emitter Coupled Logic Data Book**

**High Current Voltage Regulators**
**High Speed CMOS Data Book**
**Index to Applications Literature**
**Linear Data Book**
**Microprocessor Products Data Book**
**MOS Memory Data Book**
**Single-Chip Microcomputer**
**Small Signal Transistor Data Book**

596.   Fujitsu Microelectronics, Inc.
       3320 Scott Blvd.
       Santa Clara, CA 95051
       (408) 727-1700
       **Linear/Power Data Book**
       **Memory Data Book**
       **PROM Handbook**

597.   Intel Corp.
       3065 Bowers Ave.
       Santa Clara, CA 95051
       (408) 987-8080
       **EPROM Applications Manual**
       **EPROM Family Applications Handbook**
       - Introduction
       - EEPROM Backgrounder information
       - Applications briefs
       - EEPROM applications

       **Family Applications Handbook**
       - Introduction
       - MCS-48
       - UPI 41

       **Industrial Grade Products Data Book**
       - MCS-48
       - MCS-85
       - Peripherals
       - Special products
       - Memory Products

       **ISBC Applications Handbook**
       **Memory Components Handbook**
       - Memory overview
       - Intel memory technologies
       - Random Access Memories
       - Erasable
       - Bubble memories

       **Microcontroller Applications Handbook**
       **Microcontroller Handbook**
       - CMOS design considerations
       - 8044 architecture
       - Single component MCS-48 systems
       - 8044 applications examples

**Microprocessor Peripheral Handbook**
- Data sheets, slave processors, memory controllers data communications
- Applications notes, slave processors, memory

**OEM Systems Handbook**
- System package power
- Integrated microcomputer systems
- Memory expansion boards
- Digital I/O expansion/SGL conditioning

**Series 90 Data Catalog**
- Using the System 90
- Memory modules
- Application notes

**2920 Analog Signal Processor Design Handbook**
- Sampled data systems
- Summary of filter characteristics
- Application examples
- Design considerations
- References

598.  Mostek Corp.
      1215 Crosby Rd.
      P.O. Box 110169
      Carrollton, TX 75006
      (214) 466-6000

**Bytewyde Memory Data Book and Supplemental Data Sheets**
- Circuits/systems product guide

**Computer Products Data Book and Supplemental Data Sheets**
- Application notes
- MD series memory
- STD bus microcomputer system
- STD-Z80 BUS

**Industrial Products Data Book**
- Frequency generator
- Counter/display decoders
- Microprocessor-comp A/D converter

**Memory Data Book**
- Designers guide
- Dynamic Random Access Memory
- Pseudostatic Random Access Memory

**Micro Systems Data Book with Supplements**
- SDE series
- Development systems

**Microelectronic Data Book with Supplemental Data**
- Read Only Memory
- 68000 family
- Programmed microcomputer products

**MOSTEK Memory Systems**
**Telecommunication Data Book**
**3870/F8 Data Book**

599.   Motorola Semiconductor Products, Inc.
       5005 E. McDowell Rd.
       P.O. Box 20912
       Phoenix, AZ 85036
       (602) 962-2201
       **Bipolar PROM Cross-Reference**
       **CMOS Function Selector Guide**
       **CMOS Integrated Circuits Product Selection Guide**
       **Discrete Hybrid Components**
       - Power transistors
       - Small-signal transistors
       - Field-effect transistors
       - RF
       - Tuning diodes, switching diodes, zener diodes
       - Thyristors
       **Linear Integrated Circuits**
       - Reliability enhancement programs
       - Operational amplifiers
       - Voltage regulators
       - Application notes
       **Linear Interface Integrated Circuits**
       - Reference guide
       - Application notes/engineering bulletins
       **Linear Switchmode Voltage Regulator Handbook**
       **MECL Data Book**
       **Memory Data Manual**
       - MOS dynamic and static RAM, EPROM, EEPROM, TTL RAM, and PROM
       - Mechanical data
       **Microprocessors Data Manual**
       **Power Transistors/MOSFETS and Thyristors Data Book**
       **Rectifiers/Zener Diodes Data Book**
       **RF Data Manual**
       **Schottky TTL Data Book**
       **Small-Signal Transistor Data**

600.   National Semiconductor Corp.
       2900 Semiconductor Dr.
       Santa Clara, CA 95051
       (408) 721-5000
       **Advanced Bipolar Logic Data Book**
       **Advanced Schottky Data Book**
       **Audio/Radio Handbook**
       **CMOS Data Book**
       **Copy Microcontrollers Data Book**
       **Hybrid Products Data Book**
       **Interface/Bipolar LSI and Memory PAL Data Book**
       **Linear Applications Handbook**
       **Linear Data Book**

Logic Data Book
Memory Applications Handbook
Memory Data Book
Memory Interface Family Applications
NSC800 Microprocessor Family Handbook
Optoelectronics Handbook
Telecommunications Data Book
Transistor Data Book
Voltage Regulator Handbook

601. NEC Electronics U.S.A., Electron Division
252 Humbold Ct.
Sunnyvale, CA 94086
(408) 745-6520
Discrete/Optoelectronics Data Book
Linear Integrated Circuits Data Book

602. RCA Solid State Division
Route 202
P.O. Box 3200
Somerville, NJ 08876
(201) 685-6000
CMOS/MOS Digital Integrated Circuits
CMOS/MOS Integrated Circuits
Linear Integrated Circuits
Microboards Development Systems Software
Microprocessor/Memories/Peripherals
Power Devices

603. Signetics Corp.
811 E. Arques Ave.
Sunnyvale, CA 94086
(408) 739-7700
Analog Data Manual
Bipolar Memory Data Manual
Integrated Fuse Logic Data Manual
MOS Microprocessor Data Manual
TTL Logic Data Manual

604. Texas Instruments
13500 N. Central Expressway
P.O. Box 5012
Dallas, TX 75222
(214) 995-2011
ALS/AS Logic Circuits Data Book
Bipolar Microcomputer Components Data Book
Electro-Optical Components Data Book
High-CMOS Logic Data Book
Interface Circuits Data Book
Linear Control Circuits Data Book

**Linear/Interface Circuits Master Selection Guide**
**MOS Memory Data Book**
**9900 Family Systems Design/Data Book**
**Optoelectronics Data Book**
**Power Semiconductor Data Book**
**Transistor/Diode Data Book**
**TTL Data Book**

605. Toshiba American, Inc.
Electronics Components Division
Tustin, CA 42680
(714) 730-5000
**Audio Digital ICs Data BPPL**
**Audio IC Guide Book**
**Digital Integrated Circuits Data Book**
**Linear Integrated Circuits Data Book**
**Microwave Semiconductor Handbook**
**MOS Memory Data Book**
**Rectifier/Thyristor Data Handbook**
**RF Power Transistor Data Book**
**Transistor Data Book**

606. Westinghouse Electric Semiconductor Division
Powerex Inc.
Hillis St.
Youngwood, PA 15697
(412) 925-7272
**Power Semiconductor Data Book**
**POW-R-BLOK Data Book**
**Silicon Power Transistor Handbook**

607. Zilog
1315 Dell Ave.
Campbell, CA 95008
(408) 370-8000
**CMOS Static Rams**
**Microcomputer Components Data Book**
**Microcomputer Components Systems**

# 20    Commercial Catalog Services

The need for up-to-date product information has led to the creation of services that provide manufacturers' catalogs on a subscription basis. The most important service for electrical and electronics engineering is the *Visual Search Microfilm Files* (*VSMF*) offered by Information Handling Services (IHS), Englewood, Colorado. The one major competitor to *VSMF* is *Information Marketing International* (IMI) from Ziff-Davis, although at the time of this writing it could not match the comprehensiveness of the IHS product line. The IMI catalog service is approximately one-third the size of the IHS service and costs approximately 50 percent less. One area in which the IMI set differs quite substantially from IHS is that it includes as part of the subscription price the *D.A.T.A. Books*. Both sets have SIC (standard industrial classification), vendor indexes, and product indexes.

In both cases copies of the complete catalogs are microfilmed cover to cover and then packaged in cartridges together with any additional technical data provided by the manufacturer. This added material may be drawings, schematics, specification sheets, and applications information. In order to service specialized market segments, these catalogs may also be divided and refilmed to form a number of different products designed to meet the specific needs of engineers, specifiers, and procurement officers in particular areas. It is in this repackaging that IHS has a real advantage for electrical and electronics engineers.

Price information for both varies by type of buyer and the number of pieces bought. Because they are quite expensive, interested readers should contact a salesman for a quotation. As an example of price, the basic product catalog set from IHS costs approximately $15,000.00 per year. The following consists of special IHS products in areas of electrical and electronics engineering.

## Indexes

Each product comes with an individual index. These indexes, which come in three types, cover different areas but are organized the same way.

### 608. Brand/Trade Name Listing
Contains a list of brand and trade names with the names of the vendor who supplies that item and the location for the catalog on the cartridges.

### 609. Product/Locator Code Listing
An index of specific technical terms which directs the user to names of relevant manufacturers, and to the cartridge location for the catalogs.

### 610. Vendor Name Listing
An index of company names in alphabetical order with the vendor's divisional name structure shown, along with the primary address, telephone numbers (cable, TWX, 800, and TELEX where applicable) and page/frame numbers for locating the catalog on the cartridges.

## Product Segments

Product-oriented data segments are arranged so that like items appear side by side on the cartridges. The cartridges are updated every 120 days. The following are the most relevant for electrical and electronics work.

### 611. Communications and Components Equipment (product 141-143).
Audio and ultrasonic transducers, attenuators, and AC amplifiers.

### 612. Connectors/Terminals (product 146).
General purpose electrical connectors, coaxial connectors, pin and socket connectors, etc.

### 613. Data Processing Equipment (product 147).
Multiplexers and demultiplexers, data interfaces, computer terminals and peripherals.

### 614. Digital Integrated Circuits (product 148).
Digital gate element families, digital subsystem circuits, memory-storage circuits.

### 615. Discrete Semiconductors (product 149).
Diodes, rectifiers, bridge units, PNPN devices, transistors.

616. **Electrical Power Distribution Components** (product 150).
Electrical wiring components, power distribution assemblies, transformers/coils.

617. **Electron Tubes** (product 151).
Miniature and subminiature receiving electron tubes, standard receiving and special purpose electron tubes.

618. **Electronic Components** (product 152).
Capacitors, magnetic cores, resistors, transformers.

619. **Enclosures/Electrical Hardware** (product 153).
Chasis/boxes, circuit mounting systems, terminated circuit enclosures, insulators, test leads, patch cords interconnecting cables, power cord/cable assemblies.

# Specialized Packages

620. **Integrated Circuit Parameter Retrieval (ICPR).**
This IHS service references over 98,000 pages from 350 manufacturers and is designed for the selection and location of integrated circuits and optoelectronic devices. Access to the information is through paper indexes. Each circuit is indexed by type number, original circuit number, and function. When possible, each entry lists the data required, such as packaging, temperature range, and supplier/source. References are given to the cartridge location. The cartridges are made up of the technical pages used to extrapolate the data for the indexes. Updated every 30 days.

Also included is a *Solid State Reference Material* cartridge (cartridge R) which contains data on ICs and SCs from JEDEC Publications, JEDEC Standards, and EIA Standards.

621. **Semiconductor Parameter Retrieval (SCPR).**
References over 37,000 pages of catalogs from 180 manufacturers. Parameter data are given on transistors, thyristors, rectifiers, and diodes along with descriptions, application information, suppliers, names and cross-references to type numbers. The paper index is by type number with each entry containing a description and application of the device, suppliers names, and cartridge and frame location for a copy technical data pages used to extrapolate the data for the index.

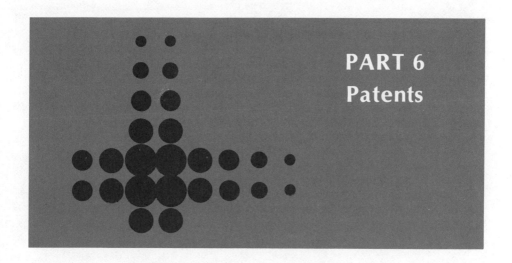

# PART 6
# Patents

# 21   Patent Literature

Patent literature, important in all technology, is particularly significant in the electrical and electronics industries where a substantial portion of designs, methods, processes, and fabrication and manufacturing techniques have resulted in patents throughout the world. By preventing unauthorized exploitation of patented inventions, governments encourage and reward new ideas.

Not only do engineers need access to what has been patented, they also need information about how to obtain protection for their own inventions. This chapter includes books designed to provide quick access to patents of interest to electrical and electronics engineers as well as books detailing the steps necessary to gain a patent in the United States.

The importance of patents is reflected by the number of bibliographic tools devoted to providing quick access to the worldwide patent literature. The following section is concerned with the print access tools; see the online section for databases devoted to patents.

622.   **Electrical Patents Index (EPI)**. London: Derwent Publishers, 1980- . weekly. full subscription $4,150.00/yr. single section $1,810.00/yr.

EPI details electrical patents in six subject categories: "Instrumentation, Measuring and Testing;" "Computing and Control;" "Semiconductors and Electronic Circuitry;" "Electronic Components;" "Communications;" and "Electronic Power Engineering." About 4,300 patent specifications are covered each week from the major countries, including the United States, United Kingdom, West Germany, Switzerland, Netherlands, Belgium, France, Canada, Soviet Union, and Sweden. An English summary and title is prepared from the full application or patent even if the original document is in English. Access to the patents is possible by patent number, patentee or assignee, title, and by Derwent assigned codes. Each section may be purchased separately.

623. **Encyclopedia of Patent Practice and Invention Management.** Edited by Robert Calvert. Huntington, N.Y.: Robert E. Kriever Publishing, c1964, 1974. 860p. index. LC 74-1028. ISBN 0-88275-184-6. $49.50.

An alphabetical arrangement of articles written by experts on various aspects of patents, including: "Claim Drafting," "Infringement," "Glossary of Patent Terms," "History of Patents," "Use of Patents," and "Profiting from Patents." Two indexes — one by case and author, the other by subject — enhance the usefulness of this book.

624. **European Patent Office (EPO) Applied Technology Series: Volume 8, Microprocessors.** Vol. 8. New York: Pergamon InfoLine, Inc., 1985. 400p. illus. index. ISBN 0-08-030575-x. $80.00.

Part of a series of patent surveys which provide authoritative overviews and detailed technical information on science and technology. The patents are condensed to feature only technical data; legal annotations are omitted. Each volume is indexed by patent number, patentee, and subject. Volume 8 is an in-depth survey of the very latest developments in microprocessors and microcomputers. Also included are applications of interest such as radiological and medical examination, printing units, vehicles, and adaption of servo-characteristics.

625. **General Information Concerning Patents: A Brief Introduction to Patent Matters.** Washington, D.C.: U.S. Department of Commerce, Patent and Trademark Office, 1984. 41p. $2.50.

This inexpensive guide to obtaining a U.S. patent is a good first step for most first time patent applicants.

626. **German Patents Gazette: Section 2 Electrical.** London: Derwent Publishers, 1968- . weekly. ISSN 0533-7534. $710.00/yr.

Weekly listing of all electrical and electronic patent applications and patents granted in West Germany. This publication serves the same purpose for German patents that the *Official Gazette* serves in the United States.

627. **Guide for Patent Draftsmen.** Washington, D.C.: GPO. (revised as necessary, last revised May 1980). c21.14/2-d. $1.50.

Contains the rules for preparing patent drawings and sample drawings illustrating the drawing conventions of the Patent Office.

628. **Invention Protection for Practicing Engineers.** By Tom Arnold. New York: Van Nostrand Reinhold, 1971. 186p. index. ISBN 0-8436-0312-7. $13.95.

A how-to guide for obtaining a U.S. patent.

629. **Official Gazette of the United States Patent and Trademark Office.** By United States Department of Commerce. Washington, D.C.: GPO, 1872- . weekly. ISSN 0098-1133. depository item or $250.00/yr.

Each weekly issue contains about 1,500 patents listed under the general categories of general and mechanical, chemical, and electrical. Each issue has an index of inventors' names and a index of patents arranged by class and subclass numbers. These weekly indexes cumulate annually into the *Index of Patents*. Each patent number is accompanied by a drawing, if relevant, and either an abstract or the largest claim of the invention. Of special interest is the listing of the names and telephone numbers for various departments of the PTO, e.g. the electrical classification group or the electrical examining group. A list of Patent Depository Libraries is also included.

630. **Patent Abstracts of Japan.** New York: Pergamon InfoLine Patents and Trademark Publications, 1977- . quarterly. $1,800.00/yr. (Section E Electrical).

Provides English language abstracts for most unexamined patent applications filed by Japanese applicants. Each issue contains approximately 500 abstracts with complete bibliographic information. Other sections available include mechanical, chemical, and physical.

631. **Patent and Trademark Tactics and Practice.** By David A. Burge. New York: Wiley-Interscience Publication, 1984. 213p. index. LC 84-2408. ISSN 0-471-80471-1. $28.95.

A highly readable book written by a registered patent attorney which focuses on the important practical points associated with development, protection, and management of patents, copyrights, and trademarks. The first chapter gives an excellent overview of the different types of intellectual property, while the second provides some good points on working with an attorney. Other important information includes the basic features of patents, for example what can be patented, how to apply, how to prosecute an application, what records the inventor needs to keep, and how to assign, license, and enforce patents and trademarks.

632. **Patent Profiles: Microelectronics-I.** Washington, D.C.: NTIS, Office of Technology Assessment and Forecasting, 1981. PB 81-179582. $7.50/paperback.

Covers patent activity in the areas of microelectronics, integrated circuits, and information processing. Includes copies of the front pages of all patents issued between January and October 1980 with information about all the previous patents cited as references.

633. **Patent Profiles: Microelectronics-II.** Washington, D.C.: NTIS, Office of Technology Assessment and Forecasting, 1983. PB 93-132613. $7.50/paperback.

The second in a series of profiles on microelectronics, this publication covers the additional areas of digital logic circuits, semiconductor memories, and speech analysis and synthesis. It includes copies of the front pages of 30 patents and briefly describes their importance. An analysis of organizational patenting patterns is also included.

634. **Patent Profiles: Telecommunications.** Washington, D.C.: NTIS, Office of Technology Assessment and Forecasting, 1984. PB 84-211044. $7.50/paperback.

Includes the area of telephony, light wave and multiplex communications, analog carrier wave communications, digital and pulse communications, television, facsimile, and telemetry. A general analysis of patenting activity in the United States is also included. A microfiche supplement is available which gives patent numbers and titles of all telecommunications patents since 1969.

635.   **Patenting Manual.** 2nd ed. By Alan M. Hale. Buffalo, N.Y.: SPI, 1983. 381p. illus. index. LC 83-061588. ISBN 0-943418-02x. $29.95.

This revised text incorporates the changes made in the 1982 Patent Law. Like other books on patenting, this covers what is patentable in the United States, the classification systems, and the procedures for obtaining a patent. Unlike other books, this one includes samples of patent drawings, claims including an in-depth discussion of claim theory, and copies of forms used by the patent office. It also includes a bibliography, a glossary of terms, and sample copies of patents.

636.   **Patents & Inventions: An Information Aid for Inventors.** Washington, D.C.: Department of Commerce, Patent and Trademark Office, 1980. 23p. GPO C21.2:P27/ 10/968. LC 64-65511. depository item or $2.50.

An introduction to patenting and patents written by the patent office.

637.   **Q & A about Patents, Answers to Questions Frequently Asked about Patents.** rev. ed. Washington, D.C.: GPO, 1982. 12p. free.

638.   **What Every Engineer Should Know about Inventing.** By G. Middendorf. New York: Marcel Dekker, 1981. 191p. index. ISBN 0-8247-1338-9. $19.95.

A good overview of the techniques for turning an idea into an invention. While some patent information is covered, the primary focus is on laboratory notebooks, experimental techniques, and ways to encourage an inventing frame of mind in engineers.

639.   **What Every Engineer Should Know about Patents.** By W. G. Konold. New York: Marcel Dekker, 1979. 211p. index. LC 79-9733. ISBN 0-8247-6805-1. $18.95.

A guide written to aid engineers obtain a U.S. patent as cheaply and efficiently as possible.

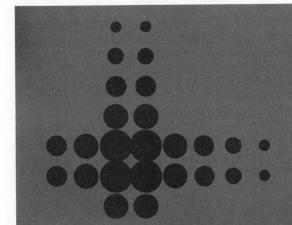

# PART 7
# Standards

# 22   Standards Literature

Standards are fundamental to many aspects of modern life including electrical and electronics engineering. These documents state how materials, products, and test methods should be defined, manufactured, installed, measured, or tested. The term standard is often used generically to cover a variety of related documents including specifications, codes of practice, recommended practices, guidelines, and nomenclature and terminology rules. Standards and specifications are formulated by companies, technical associations, professional societies, government agencies, national standards bodies, and international standards agencies. These documents are usually in effect for a limited period of time before being revised or eliminated.

Until the early part of the 20th century, standards information was uncoordinated and resided in minds and files of standards developers and users in government and industry. Corporate standards were (and still are) considered private and therefore undisclosed. Standardization activities by 1965 had grown so fragmented and widely dispersed that there was a recognized need to coordinate standards information. As a result the National Bureau of Standards, with the cooperation of industry and the national standards bodies, collected active U.S. standards. These formed the basis for a national index of voluntary standards first published in 1971 as *An Index of U.S. Voluntary Engineering Standards*.

Standards may be purchased individually or as complete sets directly from issuing agencies or vendors. (Currently there is one major vendor, Information Handling Service (IHS), which supplies complete sets of standards on microfilm or microfiche). As part of the subscription price the vendor supplies indexes that facilitate use of their products. These indexes provide access by subject and standard number. New or updated standards are supplied every 60 days. The advantages of using this service include currency, easy updating, and cross society or association indexing, that is, the ability to see if a standard on a particular topic exists without guessing which society or association promulgated it. This is important because many users assume that standards on most topics have been written either by ANSI or ASTM. This is not true. For example the *National Electrical Safety Code* is an ANSI standard while the *National Electrical Code* is from the National Fire Protection Association.

Because of the importance of standards, the following annotations include books about standards, collections, and indexes to standards as well as a list of societies and associations that produce standards and specifications which are relevant to electrical and electronics engineering.

# Standards

**640. Directory of International Regional Organizations Conducting Standards-Related Activities.** Washington, D.C.: GPO, 1983. C13.10. $8.00. (National Bureau of Standards Special Publication 649).

A directory containing information on 272 international and regional organizations which engage in standardization, certification, or other related activities. This volume describes their work as well as U.S. participants, restrictions, and the availability of English language standards.

**641. Electrical and Electronics World Standards Mutual Speedy Finder.** London: Media International Promotions, 1972. 1,500p. index. $898.00. (World Standards Mutual Speedy Finder, vol. 2).

Indexes over 18,000 electrical and electronics standards from the United States, West Germany, France, United Kingdom, and Japan. Has three indexes: standard number, key word and products-parts.

**642. Index and Directory of U.S. Industry Standards.** Edited by Information Handling Service. Englewood, Colo.: Information Handling Service, 1983. 2v. LC 81-201150. ISBN 0-89847-008-0. $425.00.

Consists of a subject index, a concordance of "old" to "new" ANSI-approved standards, a directory of standards by issuing body, and a numeric listing of standards. According to the publisher, this book represents 70 percent of the total United States standardization activity. Although the concordance of "old" to "new" ANSI-approved standards provides information that is often difficult to obtain, it's not valuable enough to justify the cost of this tool.

**643. An Index of U.S. Voluntary Engineering Standards.** Springfield, Va.: NTIS, 1971- . PB 86-154408. $13.50. (National Bureau of Standards Special Publication 681).

First published in 1971, this index was revised and updated in 1985 with the release of the new microfiche edition, *KWIC Index of U.S. Voluntary Engineering Standards.* The 1971 index was the beginning of the NBS standards database. The current index contains over 28,000 titles produced by 422 U.S. standards developing organizations. Subject access is provided for by means of a key word index which contains any significant word in the title.

**644.** **McGraw-Hill's Compilation of Data Communication Standards: Edition III Including the Omnicon Index of Standards.** Edited by Harold Folts. New York: McGraw-Hill, 1983. 1,923p. LC 86-2893. ISBN 0-07-606948-6. $275.00.

A compilation of 171 data communication, computer, and information processing standards from CCITT, ISO, ANSI, EIA, and FTSC. This is a very handy compendium of worldwide standards that are of interest to electrical and electronics engineers.

**645.** **McGraw-Hill's Electrical Code (R) Handbook.** 17th ed. Edited by J. F. McPartland. New York: McGraw-Hill, 1981. 1,133p. illus. index. LC 66-20008. ISBN 0-07-045693-3. $29.95. (Based on the Current National Electrical Code).

This reference book of commentary, discussion, and analysis of the 1981 National Electrical Code is designed to be used in conjunction with the code. The book is heavily illustrated and indexed. The index is a bit unusual in that the numbers given are not to pages in the book, but to article numbers. However, this is not a problem since the book, like the code, is organized in article order. As a result the index also can be used as an index to the code.

**646.** **National Electrical (R) Code Reference Book.** 4th ed. By J. D. Garland. Englewood Cliffs, N.J.: Prentice-Hall, 1984. 603p. illus. LC 84-2050. ISBN 0-13-609546-1. $32.95.

Based on the 1981 National Electrical Code, this book is comprised of approximately 1,500 sections and 4,000 individually treated rules. Unlike the McGraw-Hill book, the code is not covered in its entirety. Specialized sections on elevators, swimming pools, and community antenna systems are excluded.

**647.** **QRIS, Quick Reference to IEEE Standards.** Edited by Institute of Electrical and Electronics Engineers. Piscataway, N.J.: IEEE Service Center, 1980. 568p. index. LC 80-83185.

Consists of an index by key word to (IEEE and ANSI-approved IEEE) standards along with a copy of the table of contents for each standard. The tables of contents are arranged in numerical order by standard number. Somewhat dated but still valuable since the key words also include words from the title.

**648.** **Standards Activities Organizations in the United States.** Edited by Robert B. Toth. Washington, D.C.: Superintendent of Documents, 1984. 595p. index. LC 84-601084. 003-003-2602-6. C13.10. $13.00. (National Bureau of Standards Special Publication SP-681).

A directory of mandatory and voluntary standards activities in the United States. Summarizes the standardization activities of 750 organizations including federal and state agencies and 420 private sector groups that develop standards. Also includes lists of standards distributors, libraries, information centers, and union lists of standards. A subject index is included.

**649.** **Standards and Specifications Information Sources: A Guide to the Literature and to Public and Private Agencies Concerned with Technological Uniformity.** By Erasmus J. Struglia. Detroit, Mich.: Gale Research, 1965. 187p. index. LC 65-24659. ISBN 0-8103-0806-1. $48.00.

Gives short descriptions of associations, societies, and international and federal agencies that produce standards and specifications. What makes this book so useful is its discussions of nature of standards, different types of standards, and the philosophical underpinning of standardization activity.

## Selected U.S. Groups Engaged in the Production of Standards

650. **American National Standards Institute (ANSI).** 1430 Broadway, New York, New York 10018.

ANSI, the agency charged with coordinating and promoting standardization activity at the national level, is a federation of 160 technical, trade, and professional societies. ANSI's predecessor, the American Engineering Standards Committee, was started in 1918 by American Society for Testing and Materials (ASTM). In 1928 the name was changed to American Standards Association (ASA). Standards are available in paper from the association, from Information Handling Service (IHS), or from Information Marketing International (IMI) on fiche or cartridges.

651. **American Society for Testing and Materials (ASTM).** 1916 Race Street, Philadelphia, Pennsylvania 19103.

ASTM standards may be purchased either individually or as books of related standards. These books are updated yearly. Of particular interest are "Electrical Insulation Test Methods," "Electrical Insulation Specifications," and "Electronics." Also available directly, or from Information Handling Services' VSMF or IMI.

652. **Electronic Industries Association (EIA) Standards.** 2001 Eye Street Northwest, Washington, D.C. 20006.

Including JEDEC publications and quality bulletins; TEPAC publications, recommended standards, quality bulletins; JEDEC standards; and JEDEC registration. Available directly as well as from IMI or Information Handling Services' VSMF.

653. **Federal Information Processing Standards (FIPS).** U.S. Department of Commerce, National Bureau of Standards. C13.52.

Consists of the official source for federal information processing standards including automated data processing, electrical power, computer hardware, etc.

654. **Illuminating Engineering Society (IES).** 345 East 47th Street, New York, New York 10017.

Contains *Measurement and Testing Guides, Recommended Practices, IES Lighting Handbook* (see entry 231), committee reports, and energy management services. Available directly and from Information Handling Services' VSMF.

655. **Institute for Interconnecting and Packaging Electronic Circuits (IPEC).** 3451 Church Street, Evanston, Illinois 60203.

Formerly the Institute of Printed Circuits. Includes standards, the *Assembly-Joining Handbook, Test Methods Manual,* and the *Printed Wiring Design Guide.* Available directly or from Information Handling Services' VSMF.

656. **Institute of Electrical and Electronics Engineers (IEEE).** 345 East 47th Street, New York, New York 10017.

Includes definitions, standards, methods of measurement, test procedures, recommended practices, specifications, and guides. Available directly or from IMI or Information Handling Services' VSMF. The VSMF service also includes the IEEE Computer Society draft/standards and QRIS.

The standards collected in the following books are all available as part of a subscription to the IEEE Standards. They are listed separately because they are sold as separate entities and because users frequently remember the color of a book, thinking it will help them find the item. It rarely does, except in this case.

> **IEEE Brown Book: Recommended Practice for Power System Analysis.** ANSI/IEEE STD 399/1980. $24.95.

> **IEEE Buff Book: Recommended Practice for Protection and Coordination of Industrial and Commercial Systems.** IEEE STD 242/1975. $19.95.

> **IEEE Gold Book: Recommended Practice for the Design of Reliable Industrial and Commercial Power Systems.** ANSI/IEEE STD 493/1980. $19.95.

> **IEEE Gray Book: Recommended Practice for Electric Power Systems in Commercial Buildings.** IEEE STD 241/1983. $29.95.

> **IEEE Green Book: IEEE Recommended Practice for Grounding of Industrial and Commercial Power Systems.** ANSI/IEEE STD 142/1982. $17.95.

> **IEEE Orange Book: Recommended Practice for Emergency and Standby Power for Industrial and Commercial Applications.** ANSI/IEEE STD 446/1980. $19.95.

> **IEEE Red Book: Recommended Practice for Electric Power Distribution for Industrial Plants.** IEEE STD 141/1976. $19.95.

> **IEEE White Book: A Comprehensive Reference Guide on Electrical Design.** ANSI/IEEE STD 602-86. $39.95.

657. **Instrument Society of America (ISA).** P.O. Box 12277, Research Triangle Park, North Carolina 27709.

Contains *Standards and Practices for Instrumentation.* The standards are drawn from the U.S. government, Canadian Standards Association, British Standards Institute, and those written specifically for the society. Available directly and from Information Handling Services' VSMF. The IHS service also includes the most recent *Directory of Instrumentation.*

658. **National Electrical Manufacturers Associations (NEMA).** 2101 L Street Northwest, Washington, D.C. 20037.

Standards for electrical and consumer products, such as smoke detectors, capacitors, circuit breakers, etc. Available directly and from IMI or Information Handling Services' VSMF.

659. **National Fire Protection Association (NFPA).** Batter March Park, Quincy, Massachusetts 02269.

Contains the *National Electrical Code, Fire Protection Handbook, Life Safety Code, Industrial Fire Hazards, National Fire Codes, Flammable and Combustible Liquids Handbook*, and the *Standards Pamphlet Edition*. Available directly or from IMI and Information Handling Services' VSMF.

660. **Semiconductor Equipment and Materials Institute (SEMI).** 625 Ellis Street, Mountain View, California 94043.

Includes *SEMI Specifications, ASTM Test Methods,* and *MIL-Standards*. Available directly and on Information Handling Services' VSMF.

661. **Underwriters Laboratory (UL).** 333 Pfingsten Road, Northbrook, Illinois 60062.

Standards for safety that provide for construction and performance under actual and test usage of electrical apparatus and equipment. The standards are divided into electrical and nonelectrical. Available directly and from IMI or Information Handling Services' VSMF.

662. **United States Federal Standards and Specifications.** Washington, D.C.: GPO.

Consists of standards and specifications for materials and goods to be sold to or purchased by the General Services Administration. Available directly or from Information Handling Service and IMI.

663. **United States Military Standards and Specifications.** Washington, D.C.: DOD.

Consists of unclassified standards, qualified products lists, and military drawings for goods purchased by the military. Of particular usefulness are the military handbooks such as *Technical Design Standards for Frequency Division Multiplexers*. Available directly or from Information Handling Service and IMI.

# Selected List of Foreign Groups
# Engaged in Standardization

664. **British Standards Institute (BSI).** 2 Park Street, London W1A 2BS, Great Britain.

British equivalent of ANSI.

665. **Canadian Standards Association.** 178 Rexdale Boulevard, Rexdale, Ontario M9W 1R3 Canada.

Section on electrical includes, Canadian Electrical Code, switchgear assemblies, insulated conductors for power-operated electronic devices, etc.

666. **CCITT, Comite Consultatif International Telegraphique et Telephoniques.** Rue de Varembe 3, CH-1211 Geneve 20, Switzerland.

667. **DIN, Deutsches Institut fuer Normung.** Burggrafenstrasse 4-10, D-1000 Berlin 30 T:26011.

Some translations available. Individual copies are available from the Engineering Societies Library in New York and International Manufacturing and Industrial Standards Distributor.

668. **International Electrotechnical Commission (IEC).** Rue de Varembe 3, CH-1211 Geneve 20, Switzerland.

669. **International Organization for Standardization (ISO).** Rue de Varembe, CH-1211 Geneve 20, Switzerland.

670. **Japanese Industrial Standards (JIS).** 1-24, AKASKA 4, Minato-KU, Tokyo, 107 Japan.

    Limited translations available.

**PART 8**
**Other Informati**
**Sources**

# 23   Nonprint Information Sources

Many colleges and industries use audiovisual courses to train and develop students and employees. These materials may cover specific techniques and procedures or they may be more developmental in nature. Over the next few years such materials will increasingly become more important to collections serving the information needs of engineers, technicians, and designers of electrical and electronic equipment. The following is a very select list of nonprint or audiovisual materials.

671.   **Advanced Microprocessors.** Piscataway, N.J.: IEEE, 1985. $225.00.
Includes a study guide, audio cassette, and a final exam. Designed to summarize and review commercially available devices. Major emphasis is on 16- and 32-bit devices. Consists of nine lessons or approximately 80 hours.

672.   **Audio Visual Library of Computer Education.** Mill Valley, Calif.: Prismatron Productions, 1982. 15 video tapes. $1,725.00. (available in all formats).
Titles in the set include "The Computer, Understanding Computers," "Microcomputers," "The Silicon Chip," "Computer Terminals," "Secondary Storage," and "The Central Processor."

673.   **Count Any Signal.** Palo Alto, Calif.: Hewlett-Packard Co., 1980. 3/4-inch video, 10 min. $125.00.
Covers AC-DC switches and attenuator control. Illustrates settings for all types of signals.

674.   **Digital Signal Processing.** Boston: MIT Video, 1983. 22 color video tapes. $9,445.00. (available in all formats).

A course designed and taught by Alan V. Oppenheim. Tapes include "Discrete-Time Signals and Systems," "Discrete-Time Fourier Transform," "Sampling and Analyzing Frequency Response," "Z-Transform," "Representation of Linear Digital Networks," "Design of IIR Digital Filters," "Example of IIR Filter Design," and "Design of FIR Digital Filters." Each cassette may be ordered separately.

675.   **Digital Signal Processing.** Piscataway, N.J.: IEEE, 1985. $225.00.

Consists of a study guide, audio cassette, and examination. This is an advanced, 400-hour postgraduate program designed for electronics or computer engineers who are familiar with continuous linear systems theory. Optional study aids consists of a tape of 29 Fortran programs and four routines for standard 1/2-inch mainframe tape on 600-foot reel (available EBCDIC 800 bpi or 1600 bpi for $62.95) and tutorial manual, and ten floppy diskettes for an IBM-PC. Available as a total package for $329.00.

676.   **Digital Technology.** Madison, Wis.: University of Wisconsin. 24 videocassettes. $3,500.00. (available in all formats).

Emphasis is on the practical aspects of circuit analysis and design; mathematics and derivations are maximized.

677.   **Doping Silicon: Diffusion and Ion Implantation.** Ft. Collins, Colo.: Colorado State University. 8 videocassettes. $3,350.00. (available in all formats).

Presents diffusion and ion implantation as a complementary means for achieving x-y spatial patterns of doped silicon in the VLSI manufacture.

678.   **Electronic Packaging Principles.** Tucson, Ariz.: University of Arizona. 30 video-cassettes. $3,454.00. (available in all formats).

Provides an introduction to electronic packaging engineering, system requirements, system integration and fabrication, and guidelines for choice of technologies.

679.   **Future Trends in Integrated Circuit Packaging.** Tucson, Ariz.: University of Arizona. 5 videocassettes. $1,400.00. (available in all formats).

Presents an overview of the present concepts as well as future trends in packaging, device relationships, and materials compatibility.

680.   **ITTP Modules.** Research Triangle Park, N.C.: Instrument Society of America.

Modules, each of which include student workbooks, instructor's guides, and course administrator's guides, are listed below.

- **Module III Electronic Instruments.** Includes: Advance Solid-State Electronics. $4,800.00.

- **Electronic Instruments-Senors, Indicators, Transmitters.** $3,600.00.

- **Electronic Transducers.** $3,600.00.

- **Electronic Controllers.** $3,600.00.

- **Module IV Microprocessors and Digital Systems.** $7,560.00.

681.  **Spread Spectrum Signals and Systems.** Piscataway, N.J.: IEEE, 1985. $225.00.
Contains study guide, audio cassette, and final exam. Covers the concepts of spread spectrum signals and systems through eight lessons (80 hours) from basic techniques and theories to design and analysis.

682.  **Very Large Scale Integration Architecture.** Cambridge, Mass.: MIT. 39 VHS videocassettes. $12,600.
Covers the design, implementation, and use of very large-scale integrated technology and its impact on computer architecture.

# 24   Document Delivery Sources

The following is a select list of organizations which will supply copies of materials for a fee. While the charge for document delivery is much higher than that for inter-library loan, service is often faster and individuals can have materials sent directly to them. Document delivery is sometimes the only way to obtain materials not held by the major libraries.

The sources listed below include public and private research libraries as well as for-profit companies. All of the document delivery services can be contacted directly. Some will accept orders through computer search service vendors.

683.  **Chemical Abstracts**
      Dept. 31686
      P.O. Box 3012
      Columbus, OH 43210
      (800) 848-6538 ext 2823
Provides access to materials abstracted for in the last 10 years of *Chemical Abstracts.*

684. **CTIC**
P.O. Box 1210
Dublin, OH 43107
(614) 889-1310
Supplies photocopies of conference papers, technical reports, and articles.

685. **Engineering Information, Inc. (EI)**
Document Delivery Service, Room 204
345 E. 47 St.
New York, NY 10017
(800) 221-1044 or (212) 705-7301
Supplies copies of most items cited in *Engineering Index*. Rush and facsimile service available.

686. **Engineering Societies Library**
United Engineering Center
345 E. 47 St.
New York, NY 10017
(212) 705-7611
This major engineering collection is open to the public. It consists of 300,000 volumes and 5,500 subscriptions in all areas of engineering. Its primary function is to supply its membership with information and access to books, conferences, and journals. As a result members of the major engineering societies like IEEE, ASME, or ASCE may borrow materials from the library. Others may request photocopies directly or through participating libraries.

687. **Global Engineering Service**
2625 S. Hickory St.
Santa Ana, CA 92707
(800) 854-7179 or (800) 322-7013
Supplies copies of U.S. and foreign standards and specifications along with NTIS reports. Rush service is available.

688. **INFO/DOC**
BOx 17109
Dulles International Airport
Washington, DC 20041
(800) 336-0800
Supplies photocopies of articles, conference papers, NTIS and GPO documents, domestic and foreign medical literature, patents, etc. Rush service available.

689. **Information on Demand**
P.O. Box 4536
2511 Channing Way, Ste. B
Berkeley, CA 94701
(415) 841-1145
Supplies photocopies of articles, conference papers, etc.

690. **Institute for Scientific Information (ISI)**
     University City Science Center
     3501 Market St.
     Philadelphia, PA 19104
     (215) 386-0100
     Supplies articles from over 6,200 journals covered by ISI.

691. **Linda Hall Library**
     5109 Cherry St.
     Kansas City, MO 64110
     (816) 363-4600
     Open to the public. Holdings include 600,000 volumes, 866,000 microforms, and 36,000 subscriptions in all areas of science and technology. Especially strong in non-English publications.

692. **National Standards Association (NSA)**
     5161 River Rd.
     Bethesda, MD 20816
     (800) 638-8094, Telex 89-8452
     This company should not be confused with ANSI as it produces no standards; its business is to supply government and industry standards and specifications upon request. Rush service is available.

693. **Petroleum Abstracts Photoduplication Service**
     600 S. College
     Tulsa, OK 74104
     (918) 592-6000 ext 2231
     Provides photocopies of materials abstracted by *Petroleum Abstracts.*

694. **Rapid Patent Service**
     P.O. Box 2527, Eads Station
     Arlington, VA 22202
     (800) 336-5010, FAX (703) 685-3987 (Automatic), Telex 89-2362.
     Supplies copies of U.S. patents, and foreign applications and patents. A patent alerting service is available by assignee/patentee or by class/subclass.

695. **TRACOR JITCO**
     1776 E. Jefferson St.
     Rockville, MD 20852
     (301) 984-2842
     Supplies copies of journal articles, government publications, dissertations, etc.

696. **UMI Clearing House**
     300 N. Zeeb Rd.
     Ann Arbor, MI 48106
     (800) 732-0616
     Supplies photocopies of articles. Rush service is available. A sister company, UMI, will supply copies of dissertations.

697.   **Washington Bibliographic Service**
        P.O. Box 3127
        Silver Spring, MD 20901
        (301) 565-3617
   Will supply photocopies of any material available in the Washington, D.C. area.
Rush service is available.

# 25  Societies

Societies and associations exist to assist their members professionally as well as to disseminate information. The scope of activity and the amount and quality of information disseminated varies widely from society to society. Some, like ASTM, are internationally known as a result of their publications while others, like the Wire Rope Association, have a more modest audience. The amount and type of publication activity also varies. Some are primarily known for their standards, journals, or conferences, while others are known more for their professional lobbying activities.

Many meetings of interest to electrical and electronics and computer engineers are sponsored every year by societies and professional associations. These meetings serve a number of purposes. Engineers working in similar areas are able to meet and exchange information; experts in diverse fields and from the different work environments of industry, government, and academia can share their research results. A meeting with published proceedings can also provide a historical record of papers.

The large engineering societies hold an amazing number of meetings on specific topics every year. For example, the IEEE, the largest association of electrical engineers, sponsors hundreds of regional, national, and international meetings every year. Of these, over 125 are considered major and will generate a printed conference record or digest. Some sample titles are: "International Symposium on Circuits and Systems," "Computer Architecture," "Control of Power Systems," and "International Conference on Speech and Signal Processing."

Conference proceedings or transactions of meetings create difficulties for both librarians and patrons. First, many published proceedings are not included in traditional library ordering tools like *Books in Print*, but they are indexed by the major indexing sources. It is also quite common for only a limited number of copies to be available to people who did not attend the meeting. The major difficulty is that not all papers presented at meetings will appear in printed transactions or proceedings. In fact, some meetings may never have printed proceedings; rather individual papers will be distributed separately as preprints.

The difficulty in buying and accessing conferences and other society publications makes the *Encyclopedia of Associations* a very valuable tool for electrical and electronics engineering. The encyclopedia is a good place to look for addresses, short listings of publications, and telephone numbers of all types of associations from scholarly to industrial to trade groups.

The following societies regularly sponsor proceedings, conferences, symposia, and colloquiums which have printed proceedings. Some of these hold meetings regularly and libraries will want to place standing orders for their conferences.

**American Federation of Information Processing Societies (AFIPS)**
1815 N. Lynn St., Ste. 800
Arlington, VA 22209

**Association for Computing Machinery (ACM)**
11 W. 42 St. 3rd Floor
New York, NY 10036

**Electric Power Research Institute (EPRI)**
3412 Hillview Ave.
Palo Alto, CA 94304

**Electrochemical Society (ECS)**
Ten S. Main St.
Pennington, NJ 08534

**Electronic Connector Study Group (ECSG)**
P.O. Box 167
Fort Washington, PA 19034

**Institute for Interconnecting and Packaging Electronic Circuits (IPC)**
3451 Church St.
Evanston, IL 60203

**Institute of Electrical and Electronics Engineers (IEEE)**
345 E. 47 St.
New York, NY 10017

**Institution of Electrical Engineers (IEE)**
PPL Dept. 445 Hoes Ln.
Piscataway, NJ 08854

**Insulated Cable Engineers Association**
P.O. Box P
Yarmouth, MA 02664

**International Advanced Microlithography Society (IAMS)**
1635 Aeroplaza Dr.
Colorado Springs, CO 80916

**International Electronics Packaging Society (IEPS)**
P.O. Box 333
Glen Ellyn, IL 60137

**International Federation of Information Processing (IFIP)**
3 Rue du Marche
CH-1204 Geneva, Switzerland

**International Society for Hybrid Microelectronics (ISHM)**
P.O. Box 3255
Montgomery, AL 36109

**National Engineering Consortium (NEC)**
505 N. Lake Shore Dr., Ste. 4805
Chicago, IL 60611

**Semiconductor Equipment and Materials Institute (SEMI)**
625 Ellis St., Ste. 212
Mountain View, CA 94043

**Society of Automotive Engineers (SAE)**
400 Commonwealth Dr.
Warrendale, PA 15096

**Society of Photo-Optical Instrumentation Engineers (SPIE)**
1022 19th St.
P.O. Box 10
Bellingham, WA 98227

# Author/Title Index

Reference is to page and entry numbers. References to page numbers are preceded by "p." and are separated from references to entry numbers by a semicolon. Authors and titles mentioned in annotations are designated with the letter "n."

# Subject Index

Reference is to page and entry numbers. References to page numbers are preceded by "p." and are separated from entry numbers by a semicolon.